VLSI
High-Speed I/O
Circuits

- Problems, Projects, and Questions

Hongjiang Song

Copyright © 2014 by Hongjiang Song.

ISBN: 978-1-312-05875-0

All rights reserved. No part of this book may be reproduced or transmitted in any form or by any means, electronic or mechanical, including photocopying, recording, or by any information storage and retrieval system, without permission in writing from the copyright owner.

This book was printed in the United States of America.

To order additional copies of this book, contact the publisher. or copyright owner.

Preface

This book offers a collection of commonly seen problems in VLSI high-speed I/O circuit design, modeling and implementations. For beginners, this book should be used together with another published textbook in VLSI high-speed I/O circuits (VLSI High-Speed I/O Circuits – Theoretical Basis, Architecture, Modeling and Circuit Implementation. ISBN# 1-888-795-4274). This book can also be used alone for readers who already have some design experiences in VLSI high-speed I/O circuits.

This text should provide a useful reference in the form of homework, project and exam problem for my EEE598 VLSI High-Speed I/O course offered in the Engineering School at Arizona State University. To make this text more valuable, the text also includes a collection of commonly seen questions in the VLSI high-speed I/O circuit related job interviews.

Hongjiang Song, Ph.D.

Arizona State Uinversity, Arizona

Febuary 2014

CONTENTS

Preface

1

INTRODUCTION TO VLSI HIGH-SPEED I/O CIRCUITS

- VLSI High-Speed I/O Circuit Overview

- Trends of VLSI High-Speed I/O Circuits

- Key Technologies in VLSI High-Speed I/O Circuits

The demands for more bandwidth to support the increasing data rates at low cost for inter-chip and inter-system communications as the result of continuous advancements of VLSI on-chip signal processing powers and the VLSI circuit applications are among the main driving forces for the most recent developments of VLSI high-speed I/O circuits. A typical computer platform now employs multiple multi-Gbps data rate serial I/O circuits, such as the USB, PCI-Express and SATA links. Such high-speed I/O circuits are becoming standard features on the computer systems. However, this does not mean that the design challenges for faster inter-chip communication have been solved. To the contrary, it actually indicates that even higher inter-chip communication data rates yet to come and the design challenges continue. Consequently, deep understandings of the theoretical basis, the circuit operations, the circuit architectures, the modeling methodologies and the circuit implementation techniques of VLSI high-speed I/O circuit are becoming increasingly crucial for the successful developments of VLSI electronic products for lower power, higher mobility, at lower cost.

1.1 HOMEWORK AND PROJECT PROBLEMS

[1.1] Which of the following is likely the purpose for using point-to-point I/O circuit?

    a) Synchronization between parallel I/O pins.

    b) Parasitic capacitance reduction.

    c) Power saving.

    d) Cost saving.

[1.2] Which of the following is likely the reason of using equalization in high-speed I/O circuits?

    a) Skew among signal pins.

    b) Bandwidth limitation effects.

    c) ISI effect.

    d) Power saving.

[1.3] What is the main purpose of using adaptive channel equalization?

a) Power saving.

b) To compensate for time dependent channel ISI effect.

c) Synchronization among signal pins.

d) Low cost.

[1.4] Which of the following items is likely associated with the modern high-speed I/O circuits

a) Higher data rate.

b) Lower power per pin.

c) Lower power.

d) Adaptive timing control.

1.2. SAMPLE INTERVIEW QUESTIONS

1. What are the key challenges and benefits of the VLSI high-speed I/O circuits?

2. What is the Moore's Law?

3. List all VLSI high-speed I/O circuit standards you know.

4. Why serial circuit configuration is usually used in VLSI high-speed I/O circuit?

5. Why FR4, instead of more advanced material, is still commonly used as PCB material for VLSI high-speed I/O circuit?

6. What are the data rates of low, full, high, super-speed USB I/Os?

7. What are the data rates of the 1st, the 2nd, and the 3rd generation PCI-Express I/Os?

8. What are the data rates of SATA 1.0, 2.0, and 3.0 I/Os?

9. For USB3.0 I/O, why the raw throughput is 4Gbps and not 5Gbps as defined in data rate?

10. What does it mean by the full duplex and half duplex I/O?

11. List the applications of VLSI high-speed I/O circuits you know.

2

PROTOTYPE I/O CIRCUIT AND TIMING CONSTRAINT EQUATIONS

- Prototype VLSI High-Speed I/O Circuit Model

- Timing Constraint Equation

- High-Speed I/O SFG Model

A VLSI high-speed I/O circuit can be effectively modeled using the prototype circuit representation based on the basic timing elements including transmitter synchronization, a receiver synchronization, a channel and the reference clocks.

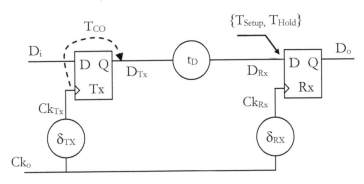

Fig.2.1 I/O prototype circuit model

The I/O delay time constraint equation provides the fundamental microscopic (bit level) delay time criteria for valid data transmission:

$$-\frac{T-T_{Setup}-T_{Hold}}{2} < \delta_{DP} + \delta_{TX} - \delta_{RX} - \frac{T-T_{Setup}+T_{Hold}}{2} < \frac{T-T_{Setup}-T_{Hold}}{2}$$

Each term in the delay time constraint equation has its very specific meaning in the practical VLSI high-speed I/O circuit practice:

- The $(\delta_{DP} + \delta_{TX} - \delta_{RX})$ term represents the effective datapath delay time of the I/O, which determines the performance of the I/O circuit;

- The $(E_{max} \equiv (T - T_{Setup} - T_{Hold})/2)$ term represents the maximum available (single-side) delay time budget for an I/O circuit at targeted data rate under the constraint of the receiver setup and hold time penalty;

- The $(\delta_{T/2} \equiv (T - T_{Setup} + T_{Hold})/2)$ term represents the eye-centering delay time shifting needed in the I/O circuit to maximize the delay time margin of the I/O circuit under delay time noise conditions;

- The $(E \equiv \delta_{DP} + \delta_{TX} - \delta_{RX} - \delta_{T/2})$ term represents the delay time (phase) tracking error of the I/O circuit, which is the I/O circuit delay time deviation from its optimal (or maximum delay time margin) operation condition.

The basic I/O circuit delay time constraint equation can be further expressed in the form of I/O delay time (or phase) tracking error equation as:

$$\begin{cases} |E| \equiv |\delta_{DP} + \delta_{TX} - \delta_{RX} - \delta_{T/2}| < E_{max} \\ \delta_{T/2} \equiv \dfrac{T - T_{Setup} + T_{Hold}}{2} \\ E_{max} \equiv \dfrac{T - T_{Setup} - T_{Hold}}{2} \end{cases}$$

Based on the I/O timing constraint equation, the prototype I/O circuit can also be modeled mathematically in phase domain using a time-domain I/O delay time tracking error signal flow graph (SFG) as:

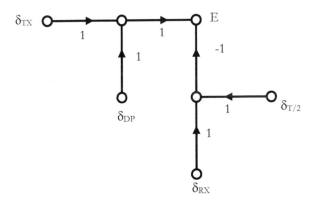

Fig. 2.2 SFG model of the VLSI high-speed I/O circuits

2.1 HOMEWORK AND PROJECT PROBLEMS

In VLSI high-speed I/O circuits, data stream is propagated in time domain. Each data bit is associated with time duration within the allowed bit time (usually one UI) where data is valid. A Timing Window as shown in figure 2.3 in static timing analysis is defined by the leading and trailing edge of the signal. Signals are assumed to be stable (valid) between the leading and trailing edge as shown in figure. Static Timing Analysis (STA) can be built using the timing window concept defined.

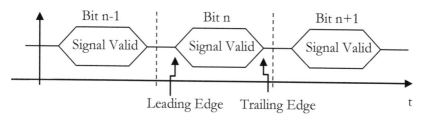

Fig.2.3 Timing window of a data bit in a data stream

[2.1] Which of the following statements about the Static Timing Analysis (STA) is generally NOT true?

a) Static Timing Analysis is a worst-case analysis.

b) Static Time Analysis is a non-simulation based approach to evaluate the propagation delay in logic circuits.

c) Static Timing Analysis needs a signal waveform for the simulation.

d) Static Timing Analysis has shorter run time than dynamic circuit simulation.

[2.2] Which of the following statements about the static timing analysis is TRUE?

a) The Require Window for a signal is defined by the setup and hold time requirements of the sampling elements.

b) The Valid Window for a signal is defined by the setup and hold time requirements of the sampling elements.

c) The Required Window for a signal is defined by the setup time requirements of the sampling elements.

d) The Valid Window for a signal is defined by the hold time requirements of the sampling elements.

[2.3] The Setup Time is ___

a) The time the signal must be stable after the sampling edge to ensure properly sampling and prevent races.

b) A clock edge on which the input data is sampled by the sampling element.

c) The latest time that the clock must remain stable.

d) The time the data signal has to be stable before the sampling edge to ensure proper sampling in the sampling element.

[2.4] The Hold Time is ____

a) The time the signal must be stable after the sampling edge to ensure properly sampling and prevent races.

b) A clock edge on which the input data is sampled by the sampling element.

c) The latest time that the clock must remain stable.

d) The time the data signal has to be stable before the sampling edge to ensure proper sampling in the sampling element.

[2.5] The setup and hold time parameters are associated with ___

a) Combinational logic circuits.

b) Sequential circuit.

c) Both combination and sequential circuits.

d) Passive VLSI circuits.

[2.6] The TCO of a flip-flop is defined as _____

    a)   Clock to data output delay of the flip-flop.

    b)   Total Capacitor at Output of the flip-flop.

    c)   Temperature Coefficient of the flip-flop.

    d)   Input to output data delay of the flip-flop.

[2.7] The Required Window's leading edge is defined by____

    a)   The earliest time the signal required being stable.

    b)   The latest time that the signal must remain stable.

    c)   The latest time the signal required to be stable.

    d)   The earliest time the signal required being table.

[2.8] The Required Windows are propagated backwards through circuits with ___

a) Both the leading and trailing edges through the maximum delays.

b) Both the leading and trailing edges through the minimum delays.

c) The leading edge through maximum delays and trailing edge through minimum delays.

d) The leading edge through the minimum delays and the trailing edge through the maximum delays.

[2.9] The Valid Widows are propagated forwards through circuit with___

a) Both the leading and trailing edges through the maximum delays.

b) Both the leading and trailing edges through the minimum delays.

c) The leading edge through maximum delays and trailing edge through minimum delays.

d) The leading edge through the minimum delays and the trailing edge through the maximum delays.

[2.10] When Required Windows are propagated backwards in the combinational circuits, the required windows usually

    a)   Become smaller.

    b)   Become bigger.

    c)   Do not change

    d)   (b) Or (c)

[2.11] When Valid Windows are propagated backwards in the combinational circuits, the required windows usually

    a)   Become smaller.

    b)   Become bigger.

    c)   Do not change

    d)   (a) Or (c)

A D-FF circuit is shown below, the setup time, hold time and the TCO of this circuit are given as:

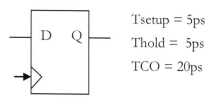

Tsetup = 5ps

Thold = 5ps

TCO = 20ps

[2.12] If a new D-FF is constructed from above D-FF with a clock buffer of delay t_d = 10ps, what is the setup time, hold time and TCO of the new D-FF?

[2.13] If a new D-FF is constructed from the given D-FF with the delay of the input buffer $t_d = 10ps$, what is the setup time, hold time and TCO of the new D-FF?

D-FF

[2.14] If a new D-FF is constructed from the given D-FF with the delay of the output buffer $t_d = 10ps$, what is the setup time, hold time and TCO of the new D-FF?

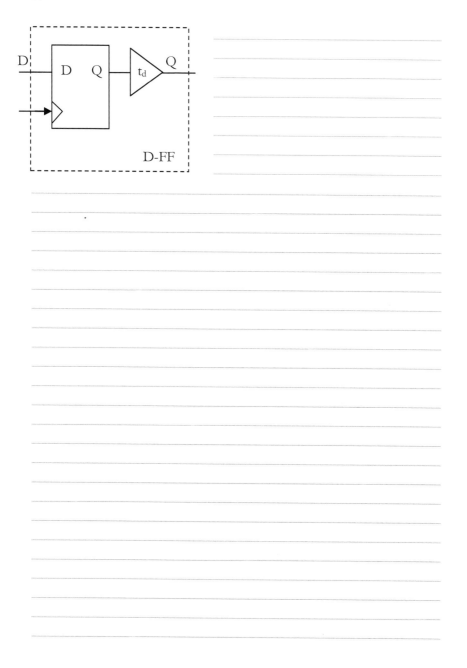

[2.15] If a new D-FF is constructed from the given D-FF with the delay of clock buffer $t_d = 10ps$, what is the setup time, hold time and TCO of the new D-FF?

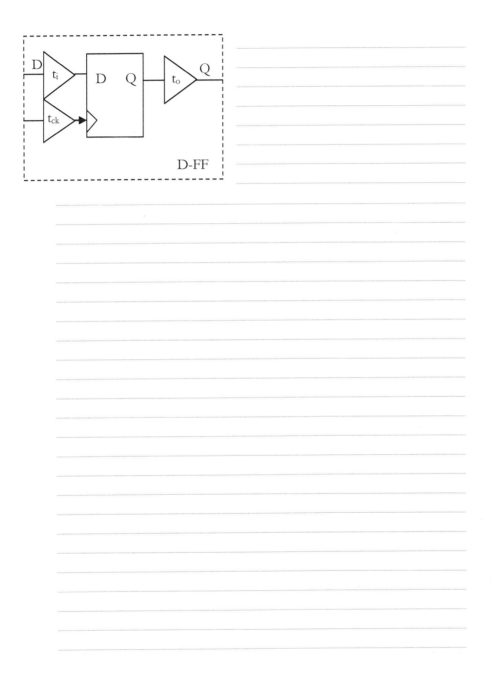

[2.16] What is the effect of the $\delta_{T/2}$ term in the high-speed I/O prototype SFG model shown in below?

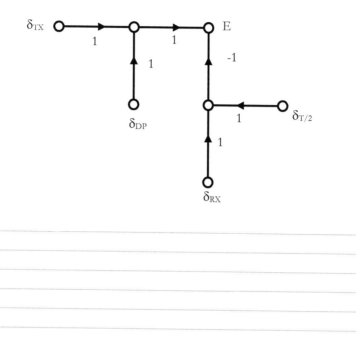

[2.17] Assuming in a VLSI high-speed I/O circuit system, the δ_{TX}, δ_{DP} and δ_{RX} can be modeled as follows. Find out the maximum data rate for this link.

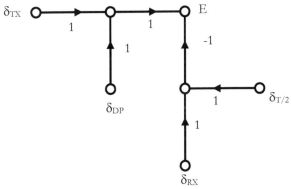

$$\begin{cases} \delta_{TX} = 50\sin(10^8 t)(ps) \\ \delta_{DP} = 20\sin(10^{10} t)(ps) \\ \delta_{RX}(s) = \dfrac{1}{1+\dfrac{s}{\omega_o}}(\delta_{TX}+\delta_{DP}) - \delta_{T/2} + 10\sin(10^9 t) \pm 25(ps) \\ \omega_o = 10^8 \, rad/s \\ t_{setup} = t_{hold} = 5\,ps \end{cases}$$

[2.18] What is the max clock frequency the circuit can operate? If T_setup= 6nS, T_hold = 2Ns, T_propagation = 10nS.

2.2. SAMPLE INTERVIEW QUESTIONS

1. What are the setup time, hold time and TCO of a VLSI synchronization circuit?

2. What determine the maximum frequency for a flip-flop-to-flip-flop path?

3. What are setup and hold time violations? How can they be eliminated?

4. Why can the hold time be neglected while calculating max frequency? Why only the setup time is considered?

5. What is capacitive loading? How does it affect slew rate of signal?

6. What does the useful-skew mean?

7. What is a false path? Give an example?

8. What are multi-cycle paths? Give example.

9. How can the power supply voltage be used to fix a timing path?

10. What is the difference between local-skew, global-skew and useful-skew?

11. What are the various timing-paths, which should be taken care in STA?

12. What is meant by virtual clock definition and why is it needed?

13. What are set up time and hold time constraints?

14. Hold time violation does not depend on clock. Is it true? If so why?

15. How power is related with clock frequency?

16. Is it possible to reduce clock skew to zero?

17. What is skew, what are problems associated with it and how to minimize it?

18. What is slack?

19. How can you increase the allowable clock frequency of a circuit?

20. What is negative slack? How it affects timing?

21. What is positive slack? How it affects timing?

22. Suppose you have a combinational circuit between two flip-flops. What will you do if the delay of the combinational circuit is greater than your clock period?

23. What is pipelining? How may it affect the performance of a design?

3

VLSI DELAY EFFECTS

- VLSI RC Delay Model

- VLSI MOS Gate Delay Models

- VLSI Interconnect Delay Models

- VLSI Transmission Line Delay Models

Electrical signals propagation through VLSI circuits will experience various time delay effects caused by the intentional or parasitic resistive, capacitive, and inductive circuit component.

Delay times of on-chip VLSI circuit elements can typically be modeled using their effective RC time constants since on-chip inductive delay effect will only show-up at very high data rate in global interconnects. The 50% delay time and the 10~90% rise time of the distributed RC-based interconnect can then be approximately expressed as:

$$t_d \approx 0.39RC = 0.39rcL^2$$

$$t_{rise} \approx 0.93RC = 0.93rcL^2$$

In most practical application, a π lumped interconnect model as shown in Fig.3.1 offers a highly simplified yet fairly accurate model for the VLSI RC interconnect. Under the π model, a VLSI distributed interconnect circuit can be modeled as a lumped resistor with the resistance equals to the total interconnect parasitic resistance and two identical lumped capacitors with capacitance equals to half of the total parasitic interconnect capacitance.

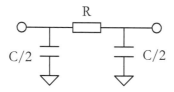

Fig.3.1 The π lumped VLSI interconnect model

The 50% swing delay and 10% to 90% rise time of interconnect under the π model can then be expressed as:

$$t_d \approx 0.7 \cdot 0.5RC = 0.35rcL^2$$

$$t_{rise} \approx 2.2 \cdot 0.5RC = 1.1rcL^2$$

For complicated RC networks including multiple segments and branches, the Elmore delay model is commonly used model for the delay and rise time calculation.

For the VLSI high-speed I/O circuits, the off-chip circuit element, such as the channels, including the package parasitic, the connectors, the PCB trace or the cable, are among the dominant delay effects. High-speed I/O circuit channels are typically being modeled using the lossless or lossy transmission line circuits combined with lumped RLC components contributed from package and trace discontinuities.

Transmission line effects, such as the reflection, ringing, cross-talk, ISI and attenuation are among key factors that impact the performances of the VLSI high-speed I/O circuits.

The time-domain response of the lossless transmission line can be analyzed based on the injection, propagation and reflections diagram through the superposition method.

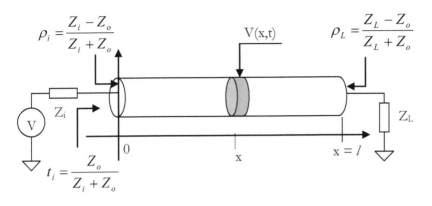

Fig.3.2 Transmission line system parameters

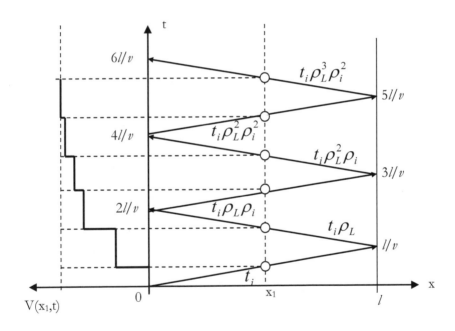

Fig.3.3 Reflection diagram of lossless transmission line

3.1 HOMEWORK AND PROJECT PROBLEMS

[3.1] A simple on-chip RC based VLSI circuit delay model is given in the figure below. Where the ideal buffer is assumed to have zero delay and parasitic capacitance respectively. It also has infinite gain and is clamped to 0 and Vcc.

Derive the delay time td in terms of circuit parameter R and C if the threshold of the buffer is Vcc/2.

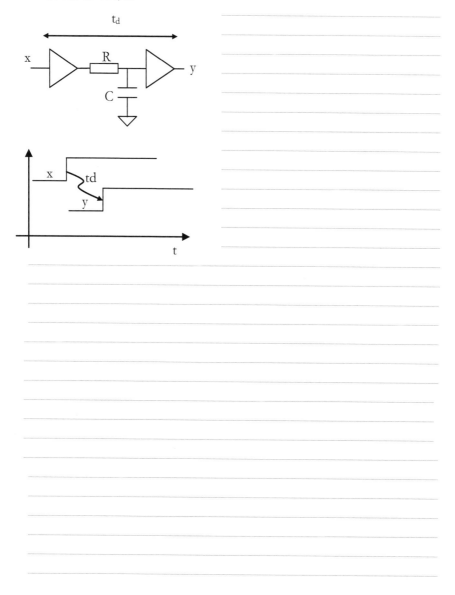

[3.2] For the VLSI circuit delay model in problem [3.1], derive the delay time td in terms of circuit parameter R, C and k if the threshold of the buffer is kVcc.

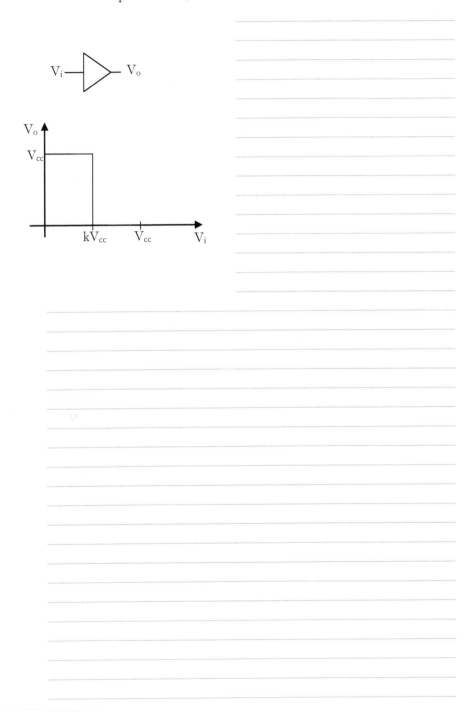

[3.3] For problem [3.1], find the delay for the signal falling edge (i.e. input signal transition from Vcc to 0).

[3.4] For problem [3.3], find the delay time of the signal falling edge.

[3.5] For the circuit shown in figure below, find the signal rise and falling delay times. Assuming the buffer has zero delay with infinite gain and clamped to 0 and Vcc. Ignoring the forward voltage drop of the diode and the parasitic capacitance and resistance effects of the diode.

[3.6] For the circuit shown in figure below, find the signal rise and falling transition delay times in terms of R1, R2, and C. Assuming the buffer has zero delay with infinite gain and clamped to 0 and Vcc. The threshold of the buffer is Vcc/2. Ignoring the forward voltage drop of the diode and the parasitic capacitance and resistance effects of the diode.

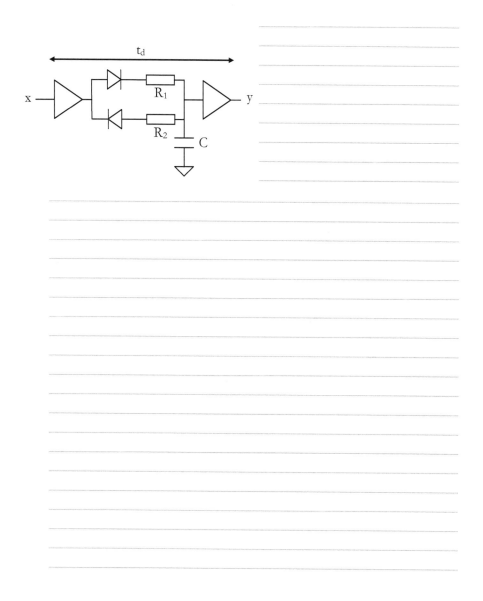

[3.7] For the circuit shown in figure below, find the signal rise and falling transition delay times for the buffer threshold of kVcc in terms of R1, R2, C and k. Assuming the buffer has zero delay with infinite gain and clamped to 0 and Vcc. Ignoring the forward voltage drop of the diode and the parasitic capacitance and resistance effects of the diode.

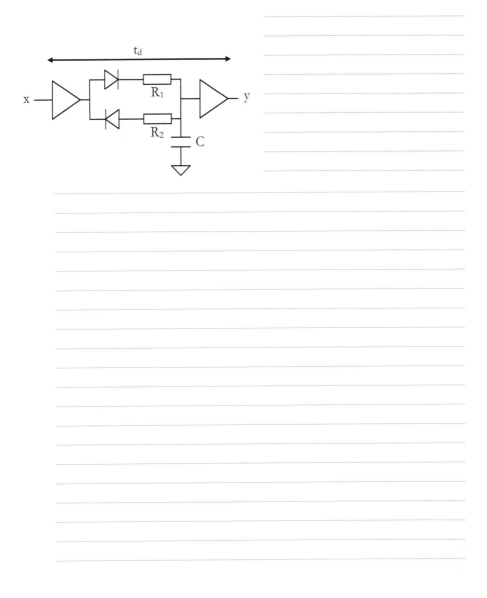

[3.8] If the effective drive resistance of the MOS device can be approximated modeled by the channel resistance at the Vds = 0 and Vgs = Vcc operation condition (i.e. linear buffer model), derive the effective pull-up (Rp) and pull-down (Rn) resistances of the given CMOS buffer in terms of βp, βn, Vtn, Vtp, and the Vcc.

[3.9] Derive the comparison threshold Vth of the given CMOS buffer in terms of βp, βn, Vtn, Vtp, and the Vcc.

[3.10] If the effective drive resistance of the MOS device can be approximated modeled by the channel resistance at the Vds = Vcc/4 and Vgs = Vcc operation condition (i.e. modified linear buffer model), derive the effective pull-up (Rp) and pull-down (Rn) resistances of the given CMOS buffer in terms of βp, βn, Vtn, Vtp, and the Vcc.

[3.11] If the effective drive resistance of the MOS device can be approximated modeled by the channel resistance at the Vds = Vcc/2 and Vgs = Vcc/2 operation condition (i.e. saturation based buffer model), derive the effective pull-up (Rp) and pull-down (Rn) resistances of the given CMOS buffer in terms of βp, βn, Vtn, Vtp, and the Vcc.

[3.12] Derive the FO-of-N delay time of the given VLSI CMOS circuit for step input under the linear buffer model in terms of β ($\beta=\beta p=\beta n$), Vt (Vt =Vtn=|Vtp|), Vcc, L (length of the MOS gate), N (fan-out), and Cox (CMOS gate capacitance per unit area). Ignore the parasitic drain capacitance of the device.

[3.13] Derive the FO-of-N rise/fall delay time of the given VLSI CMOS circuit under the linear buffer model in terms of βp, βn, Vtn, Vtp, Vcc, L (length of the MOS gate), N (fan-out), and Cox (CMOS gate capacitance per unit area). Ignore the parasitic drain capacitance of the device.

[3.14] Derive the FO-of-N fall/rise delay time of the given VLSI CMOS circuit under the linear buffer model in terms of βp, βn, V_{tn}, V_{tp}, V_{cc}, L (length of the MOS gate), N (fan-out), and C_{ox} (CMOS gate capacitance per unit area). Ignore the parasitic drain capacitance of the device.

[3.15] Derive the rise/fall delay time of the given VLSI CMOS circuit under the linear buffer model in terms of $\beta=\beta p=\beta n$, $Vt=Vtn=Vtp$, Vcc, L (length of the MOS gate), N (fan-out), and Cox (CMOS gate capacitance per unit area). Ignore the parasitic drain capacitance of the devices.

[3.16] Derive the rise/fall delay time of the given VLSI CMOS circuit under the linear buffer model in terms of $\beta=\beta_p=\beta_n$, $V_t=V_{tn}=V_{tp}$, Vcc, L (length of the MOS gate), N (fan-out), and Cox (CMOS gate capacitance per unit area). Ignore the parasitic drain capacitance of the devices.

[3.17] Derive the fall/rise delay time of the given VLSI CMOS circuit under the linear buffer model in terms of $\beta=\beta_p=\beta_n$, $V_t=V_{tn}=V_{tp}$, Vcc, L (length of the MOS gate), N (fan-out), and Cox (CMOS gate capacitance per unit area). Ignore the parasitic drain capacitance of the devices.

[3.18] Derive the fall/rise delay time of the given VLSI CMOS circuit under the linear buffer model in terms of $\beta=\beta_p=\beta_n$, $V_t=V_{tn}=V_{tp}$, V_{cc}, L (length of the MOS gate), N (fan-out), and C_{ox} (CMOS gate capacitance per unit area). Ignore the parasitic drain capacitance of the devices.

[3.19] Derive the rise/fall delay time of the given VLSI CMOS circuit under the linear buffer model in terms of βp, βn, Vtn, Vtp, Vcc, L (length of the MOS gate), N (fan-out), and Cox (CMOS gate capacitance per unit area). Ignore the parasitic drain capacitance of the devices.

[3.20] Derive the rise/fall delay time of the given VLSI CMOS circuit under the linear buffer model in terms of βp, βn, Vtn, Vtp, Vcc, L (length of the MOS gate), N (fan-out), and Cox (CMOS gate capacitance per unit area). Ignore the parasitic drain capacitance of the devices.

[3.21] Derive the fall/rise delay time of the given VLSI CMOS circuit under the linear buffer model in terms of β_p, β_n, V_{tn}, V_{tp}, V_{cc}, L (length of the MOS gate), N (fan-out), and Cox (CMOS gate capacitance per unit area). Ignore the parasitic drain capacitance of the devices.

[3.22] Derive the fall/rise delay time of the given VLSI CMOS circuit under the linear buffer model in terms of β_p, β_n, Vtn, Vtp, Vcc, L (length of the MOS gate), N (fan-out), and Cox (CMOS gate capacitance per unit area). Ignore the parasitic drain capacitance of the devices.

[3.23] In the empirical based behavior CMOS buffer model, MOS buffer delay versus FO (N) can be simulated and fitted to the linear model. For a given 0.18um VLSI process technology, simulate the rise/fall delay time of the given circuit and fit to find the parameter Arise/fall and Brise for (W/L)n = 0.2/0.18 and (W/L)p = 0.24/0.18.

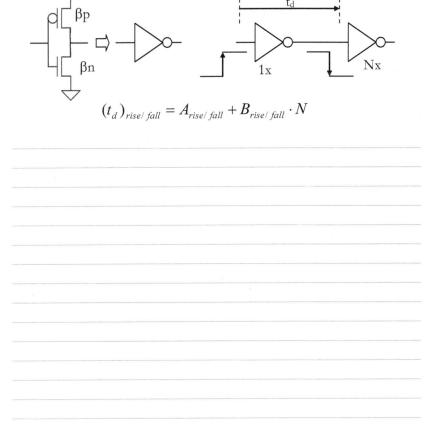

$$(t_d)_{rise/fall} = A_{rise/fall} + B_{rise/fall} \cdot N$$

[3.24] In the empirical based behavior CMOS buffer model, MOS buffer delay versus FO (N) can be simulated and fitted to the linear model. For a given 0.18um VLSI process technology, simulate the fall/rise delay time of the given circuit and fit to find the parameter Arise/fall and Brise for (W/L)n = 0.2/0.18 and (W/L)p = 0.24/0.18.

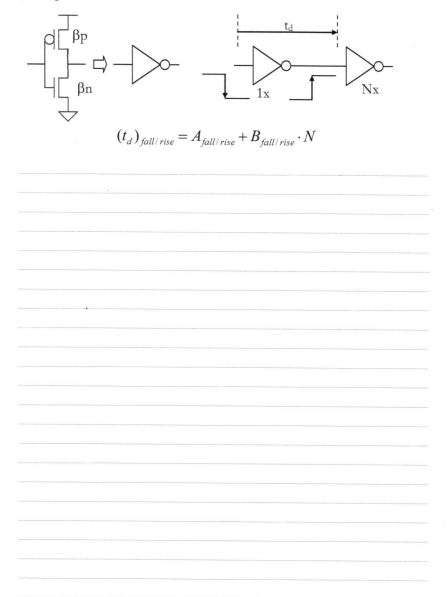

$$\left(t_d\right)_{fall/rise} = A_{fall/rise} + B_{fall/rise} \cdot N$$

[3.25] By combining (averaging) the parameters from [3.23] and [3.24] we can get a simplified buffer behavioral model between the inverter delay versus the FO (N).

$$t_d = A + B \cdot N$$

Find the criteria of the minimal delay FO value Nmin for repeater insertion in terms of A and B that yield shorter delay. Assuming signal polarity change is allowed.

[3.26] For the average delay versus FO model given above, find the criteria of the minimal delay FO value Nmin for repeater insertion in terms of A and B that yield shorter delay and keep the same circuit function. (i.e. no signal polarity change is allowed).

[3.27] For the average delay versus FO model given above, design a minimal delay repeater chain (number of buffer stage and FO of each stage) circuit for the VLSI circuit given below.

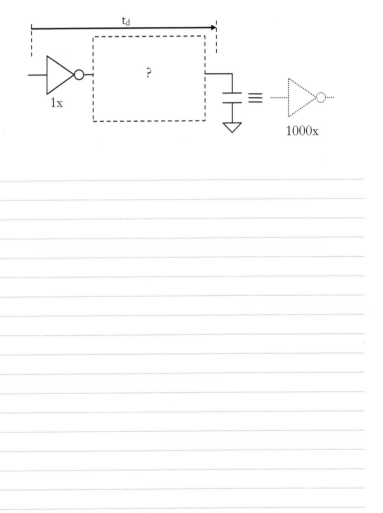

[3.28] Repeat [3.23], [3.24], and [3.25] for the following VLSI NAND based circuit to get a simplified buffer behavioral model between the NAMD gate delay versus the FO (N).

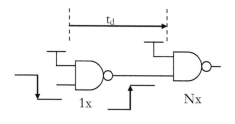

$$t_d = A + B \cdot N$$

[3.29]   Repeat [3.26] and [3.27] for above NAND gate chains.

[3.30] A normalized VLSI interconnect has a unity resistance and capacitance uniformly distributed along the interconnect.

$$\{R, C\} = \{1, 1\}$$

Calculate and plot the normalized Elmore delay (the output reach 50% of the full swing for the step input) of the interconnect versus N, where the interconnect is divided into N identical segments. Please use simple RC model shown below for each interconnect segment.

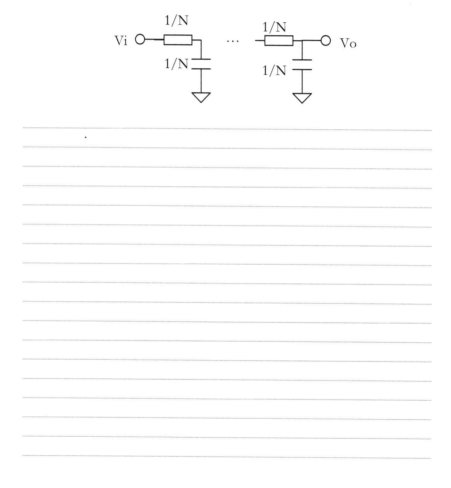

[3.31]   A normalized VLSI interconnect has a unity resistance and capacitance uniformly distributed along the interconnect.

$$\{R, C\} = \{1, 1\}$$

Calculate and plot the normalized Elmore delay (the output reach 50% of the full swing for the step input) of the interconnect versus N, where the interconnect is divided into N identical segments. Please use simple RC model shown below for each interconnect segment.

[3.32]   A normalized VLSI interconnect has a unity resistance and capacitance uniformly distributed along the interconnect.

$$\{R, C\} = \{1, 1\}$$

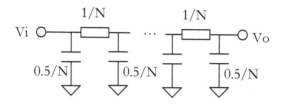

Calculate and plot the normalized Elmore delay (the output reach 50% of the full swing for the step input) of the interconnect versus N, where the interconnect is divided into N identical segments. Please use Π- RC model for each interconnect segment.

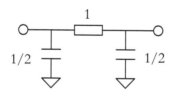

Compare the results obtained from [3.30], [3.31] and [3.32] with the π model below:

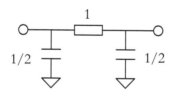

[3.33] In a VLSI circuit layout, the distance (i.e. L) and the total layout area S of the interconnect are pre-defined (fixed). Please determine function y that yields minimal signal delay time. Assuming the interconnect has the normalized sheet resistance (Rs=1) and unity area capacitance (C =1).

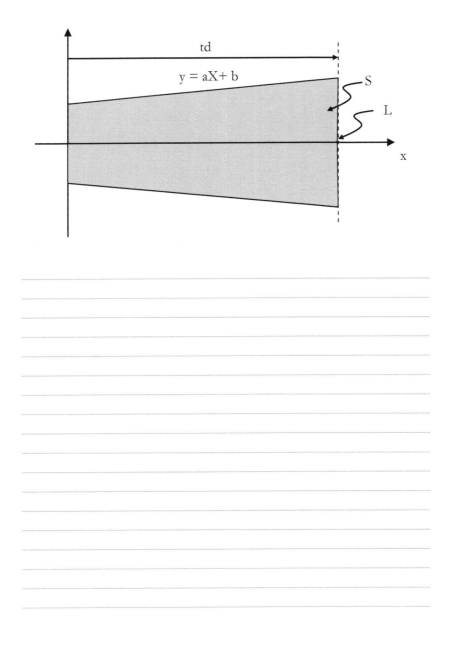

[3.34] In a VLSI circuit layout in 0.18um CMOS technology given in the appendix, two inverters of given sizes are 10,000um apart (as shown in figure). Design a repeater circuit (including both the CMOS gates and interconnect placement) such that the delay time td (average of rise and fall transitions) is minimized.

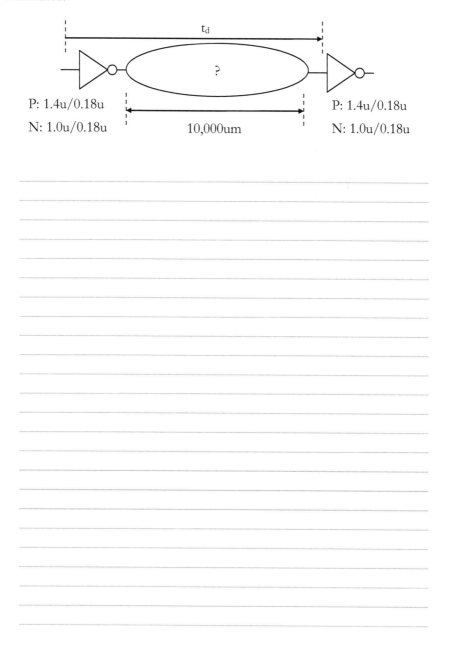

[3.35] A lumped LRC VLSI interconnect circuit model is shown in figure below. Derive the relation among parameter R, L, C such that no overshoot will be observed at the output when there is a step input signal.

[3.36] A lumped LRC VLSI interconnect circuit model is shown in figure below. Derive the relation among parameter R, L, C such that no overshoot will be observed at the output when there is a step input signal.

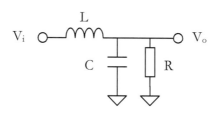

[3.37] For the lumped LRC VLSI interconnect circuit model shown below and the step input signal, simulate and comment for the 50% delay time (i.e. the output reach 50% of the full swing) in terms of parameter $T=(LC)^{0.5}$ and $Q=(L/C)^{0.5}/R$.

[3.38] For the lumped LRC VLSI interconnect circuit model shown below and the step input signal, simulate and comment for the 50% delay time (i.e. the output reach 50% of the full swing) in terms of parameter $T=(LC)^{0.5}$ and $Q=(L/C)^{0.5}/R$.

[3.39] A normalized VLSI interconnect has a unity inductance 1 and capacitance 1 uniformly distributed along the interconnect.

$$\{L, C\} = \{1, 1\}$$

$V_i(t)$ ⎯ Rt ⎯ [   ] $V_o(t)$

Calculate the source impedance Rt that is required to terminate the line (assuming the far end of the line is open).

[3.40] A normalized VLSI interconnect has a unity inductance 1 and capacitance 1 uniformly distributed along the interconnect.

$$\{L, C\} = \{1, 1\}$$

Calculate the load impedance Rt that is required to terminate the line (assuming the far end of the line is open).

[3.41] Use the reflection diagram to find the output waveform of the step input to the following transmission line circuit.

[3.42]   Use the reflection diagram to find the output waveform of the step input to the following transmission line circuit.

[3.43]   Use the reflection diagram to find the output waveform of the step input to the following transmission line circuit.

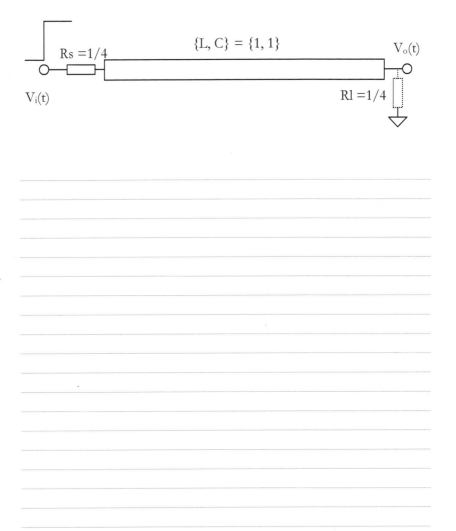

[3.44]   Use the reflection diagram to find the output waveform of the step input to the following transmission line circuit.

[3.45] Use the reflection diagram to find the output waveform of the step input to the following transmission line circuit.

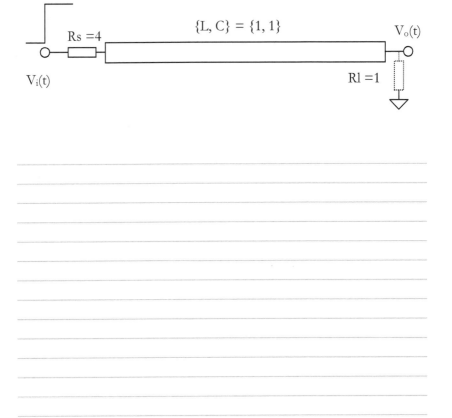

[3.46]   Use the reflection diagram to find the output waveform of the step input to the following transmission line circuit.

[3.47] Write the Elmore delay from Vi to Vo in terms of the resistors and capacitors in the circuit.

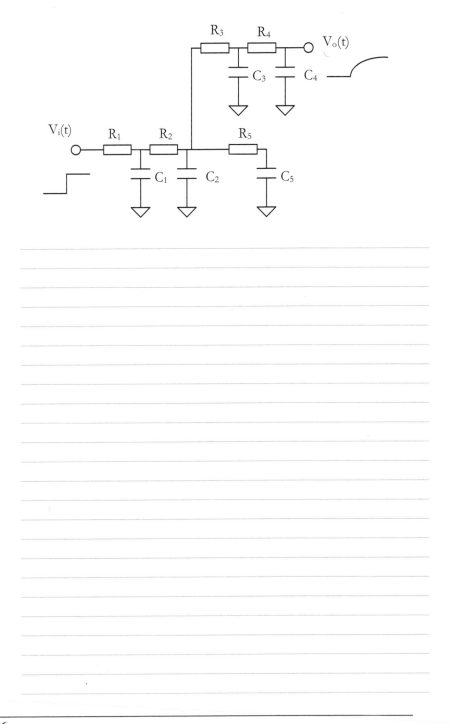

[3.48] Find the delay time td in terms of equivalent R and C parameters in the following circuit.

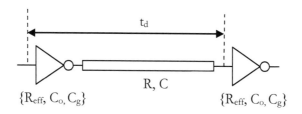

$$t_d$$

$$R, C$$

$$\{R_{eff}, C_o, C_g\} \qquad \{R_{eff}, C_o, C_g\}$$

[3.49] Generate the reflection diagram for V x in the following normalized transmission line circuit.

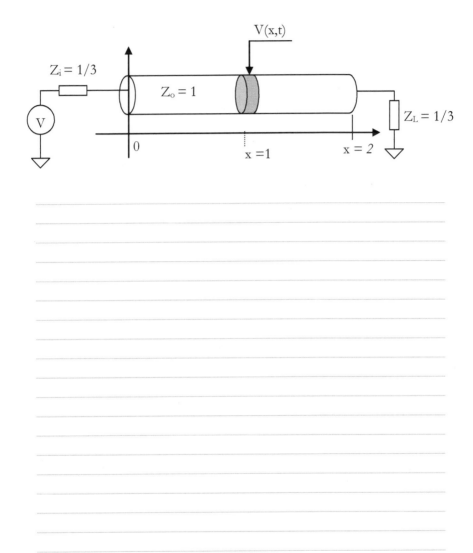

[3.50] Shown in figure is a normalized buffer delay versus fanout plot. Determine the normalized buffer delay of such circuit with FO = 1000.

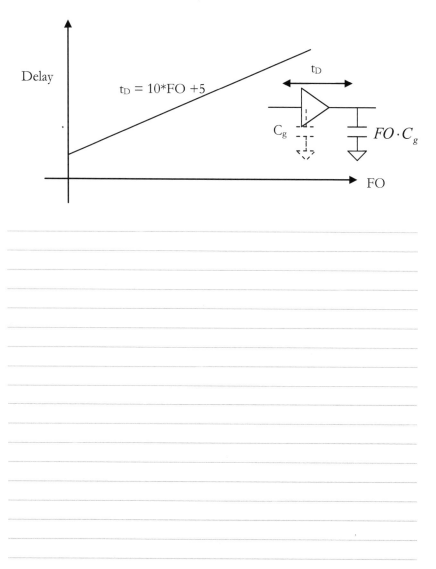

[3.51] For shown in above problem, determine the number of (buffer) repeater stages to minimize the normalized buffer delay of such circuit with FO = 1000.

[3.52] For shown in above problem, determine the minimal buffer delay of such circuit with FO = 1000 and with buffer repeaters.

[3.53] The delay time of normalized interconnect circuit is shown in figure below. Calculate the normalized delay time of such interconnect for $L = 100$.

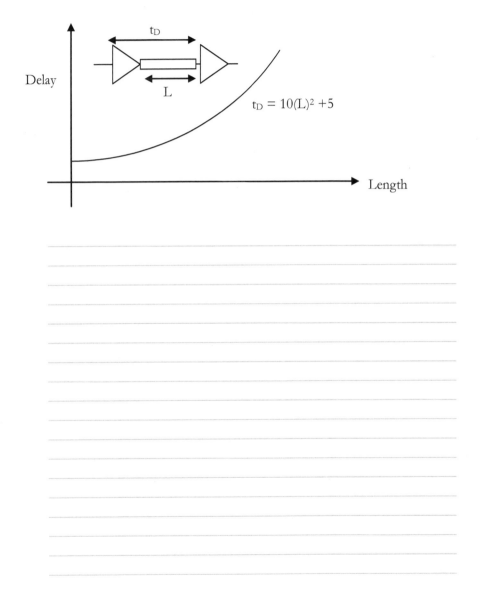

$t_D = 10(L)^2 + 5$

[3.54] For the interconnect circuit given in above problem, determine the number of repeater stages that minimizes the circuit delay. What is the minimal circuit delay with the repeaters?

3.2 SAMPLE INTERVIEW QUESTIONS

1. What is the effect of the β parameter of MOS device on circuit delay?

2. Draw typical Vds-Ids curve for a MOS device. How it varies with a) increasing Vgs b) velocity saturation c) channel length modulation d) W/L ratio.

3. What is the body effect? How does it impact the delay time of the circuit?

4. How does the load capacitance impact the delay of a CMOS circuit?

5. Why is NAND gate usually preferred over NOR gate in VLSI circuits?

6. What are commonly used sizing optimization approaches of the CMOS inverter?

7. What happens to delay if the layout parasitic resistance CMOS device is included?

8. What are the limiting factors to use the power supply to reduce delay?

9. How does resistance of the metal lines vary with increasing thickness and increasing length?

10. In the I-V characteristics curve of MOS device, why is the saturation curve flat or constant?

11. What are the different regions of operation in a CMOS transistor?

12. What are the effects of the output characteristics for a change of the beta (β) value?

13. What is the effect of MOS body bias?

14. Explain the operation of a CMOS inverter and its characteristics?

15. Draw the layout of a CMOS inverter.

16. Explain the origin of the various capacitances in the CMOS transistor and the physical reasoning behind it.

17. Why should the number of CMOS transistors that are connected in series be reduced?

18. What is an interconnect repeater?

19. What is Elmore delay algorithm?

20. What is the effect of supply voltage Vdd on delay?

21. What is the effect of load capacitance on delay, rise and fall times?

22. What is floor plan? How will floorplan impact the delay of a signal?

23. What happens if we increase the number of contacts or vias from one metal layer to the next?

24. Explain the various MOS device capacitances and their significance.

25. Let A & B be two inputs of the NAND gate. If signal A arrives at the NAND gate later than signal B. To optimize delay, of the two series NMOS inputs A & B, which one would you place near the output?

26. Why should we have a balanced clock tree?

27. What is the difference between two- and three-dimensional analysis of interconnect capacitance.

28. Guard bands are usually built into the timing estimates employed by logic synthesis, cell placers, and other CAD tools. What is the penalty when the guard bands are too large?

29. What could be gained if the timing estimates could be made more accurate?

30. The capacitance on a node is the sum of several components. What is meant by fringe capacitance?

31. How does reducing the width of a conductor affect the fringe capacitance?

32. How can the parasitic capacitance between two signal nodes possibly cause the signal transition on one of the nodes to be unexpectedly speed up?

33. How can a layout designer help ensure that the propagation delay along two conductors is very similar?

34. What are the cell delay and the net delay?

35. What are delay models and difference between them?

36. What is wire load model?

37. How is the problem of driving a clock node different from that of designing a regular signal node?

38. Draw the stick diagram of a 2-input NAND gate.

39. For CMOS logic, give the various techniques to minimize power consumption.

40. Why don't we use just one NMOS or PMOS transistor as a transmission gate?

41. For a NMOS transistor acting as a pass transistor, if the gate is connected to VDD, gives the output for a square pulse input going from 0 to VDD.

42. While trying to drive a large load, driver circuits are designed with number of stages with a gradual increase in sizes. Why is this done so? Why not use just one big driver gate?

43. What is the effect of increase in the number of contacts and vias in the interconnect layers?

44. How does the resistance of the metal layer vary with increasing thickness and increasing length?

45. What are set up time & hold time constraints? What do they signify? Which one is critical for estimating maximum clock frequency of a circuit?

46. Suppose you have a combinational circuit between two flip-flops driven by a clock. What will you do if the delay of the combinational circuit is greater than your clock signal? (You can't resize the combinational circuit transistors).

47. The answer to the above question is breaking the combinational circuit and pipelining it. What will be affected if you do this?

48. How much is the max fan out of a typical CMOS gate? What are the limiting factors?

49. What are the measurements to be taken to design for optimized area?

50. What are dynamic logic gates? What are their advantages over conventional logic gates?

51. How can we reduce the power consumption for CMOS logic?

52. What are some of the major techniques that are usually considered when one wants to speed up the propagation speed of a signal?

53. What are the major factors that determine the speed that a logic signal propagates from the input of one gate to the input of the next driven gate in the signal path?

4

VLSI DELAY VARIATION EFFECTS

- Basic VLSI Circuit Delay Effects
- VLSI Time Constant Modulation Effects
- VLSI Noise to Jitter Transformation
- ISI Effects

Electrical signals in VLSI high-speed I/O circuits suffer from two major types of delay variation effects, including the time-dependent and time-independent delay variations. Time-independent delay variations are caused by VLSI device mismatches, PVT and application condition (such as channel trace length and loading conditions) variations. Time-independent delay variations result in delay skew effects in the VLSI high-speed I/O circuits. The time-dependent delay variations, on the other hand, are usually caused by various noise effects that result in the jitter.

VLSI circuit delay variations can be modeled using three key mechanisms, including the time constant modulation, the voltage noise to jitter conversion, and the inter-symbol interference (ISI) effects. Time constant modulation effects are caused by the dependency of VLSI circuit time constant on the PVT and application conditions. The voltage noise to jitter conversion effects are voltage noise (such as the device thermal noise, power supply noise, substrate noise, cross-talk, reflection and ringing) induced delay variations. Voltage noise to jitter conversion effects can be modeled using the voltage noise to jitter transfer functions. The ISI induced delay variation effects as caused by the pattern or duty-cycle dependent signal delay variations that are mainly due to frequency dependent response of VLSI circuits, such as the bandwidth limitation effects of the devices and channel.

VLSI RC circuit delay variation effects can be explained using the simple circuit shown in Fig.4.1 with a step signal propagating through the simple RC network.

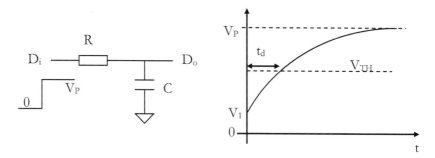

Fig.4.1 Basic VLSI circuit delay uncertainty model

The output signal waveform of such RC circuit for step input signal can be solved analytically as:

$$V_o(t) = V_P - (V_P - V_1)e^{-\frac{t}{RC}} + V_N(t)$$

Where V_P and V_1 are the peak voltage swings of the step input signal and the initial voltage of the output signal respectively. V_N is the equivalent voltage noise floor contributed from voltage noise sources and dc offset. The delay time t_d with respect to a given voltage threshold V_{TH} is given as:

$$t_d = (RC)\ln(\frac{V_P - V_1}{V_P + (V_N - V_{TH})}) = (RC)\ln(\frac{1 - V_1/V_P}{1 + \frac{V_N - V_{TH}}{V_P}})$$

It can be seen from above equation that there are three major circuit non-ideal effects that contribute to the circuit delay time variations:

- Circuit RC time constant variation. Such type of variation is resulted from the PVT variation and it shows as delay skew effect. This delay variation term may also include duty-cycle distortion effects due to the rise and fall times mismatch and the jitter effects due to the variation of the time constant in time.

- The $(V_N-V_{TH})/V_P$ term represents the noise to jitter conversion effects in the VLSI circuit. It can be seen that either the voltage noise or the threshold variation can result in the delay variation effect. Such delay time uncertainty effect can also introduce the skew in the circuit, if there are a static offset in either the signal or threshold circuits.

- The V_1/V_P term represents the ISI effect that is resulted from the voltage initial condition variation impacted due to the settling of the signal beyond its bit time. Such effect is usually resulted from the bandwidth limitation of the VLSI circuits or the high-speed I/O channel.

4.1 HOMEWORK AND PROJECT PROBLEMS

[4.1]-[4.8] Three basic jitter injection mechanisms include (a) time constant modulation, (b) noise to jitter transfer function, (c) ISI effects. Which type of the jitter injection mechanisms is related to the following VLSI circuit jitter?

[4.1] CMOS gate drive strength dependency on the supply voltage.

[4.2] Cross-talk noise.

[4.3] Substrate-coupling noise.

[4.4] Bandwidth limitation effect of the VLSI interconnects.

[4.5] Ground bounce effects.

[4.6] Pattern dependent jitter effects.

[4.7] Un-matched I/O channel.

[4.8] High-Q LRC network.

[4.9] A simple RC based VLSI circuit jitter injection model is given in the figure below. Where the ideal buffer is assumed to have zero delay and parasitic capacitance. It has infinite gain and is clamped to 0 and Vcc.

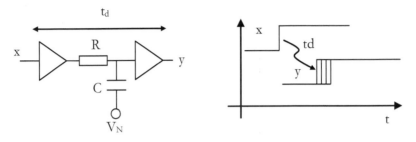

Assuming the threshold of the buffer is Vcc/2 and $V_N << V_{cc}$. Derive the cross-talk noise to jitter transfer function $d(t_d)/d(V_N)$ in terms of circuit parameter R, C, and V_{CC}.

[4.10] A simple RC based VLSI circuit jitter injection model is given in the figure below. Where the ideal buffer is assumed to have zero delay and parasitic capacitance. It has infinite gain and is clamped to 0 and Vcc.

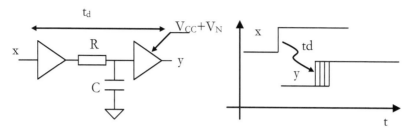

Assuming the threshold of the buffer is Vcc/2 and $V_N \ll V_{CC}$. Derive the power supply noise to jitter transfer function $d(td)/d(V_N)$ in terms of circuit parameter R, C, and V_{CC}.

[4.11] A simple RC based VLSI circuit jitter injection model is given in the figure below. Where the ideal buffer is assumed to have zero delay and parasitic capacitance. It has infinite gain and is clamped to 0 and Vcc.

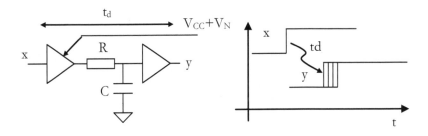

Assuming the threshold of the buffer is Vcc/2 and $V_N \ll Vcc$. Derive the power supply noise to jitter transfer function $d(td)/d(V_N)$ in terms of circuit parameter R, C, and V_{CC}.

[4.12] A simple RC based VLSI circuit jitter injection model is given in the figure below. Where the ideal buffer is assumed to have zero delay and parasitic capacitance. It has infinite gain and is clamped to 0 and Vcc.

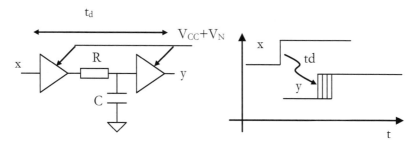

Assuming the threshold of the buffer is Vcc/2 and $V_N \ll V_{cc}$. Derive the power supply noise to jitter transfer function $d(td)/d(V_N)$ in terms of circuit parameter R, C, and V_{CC}.

[4.13] A simple RC based VLSI circuit jitter injection model is given in the figure below. Where the ideal buffer is assumed to have zero delay and parasitic capacitance. It has infinite gain and is clamped to 0 and Vcc.

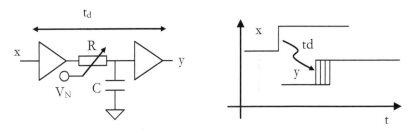

Assuming the threshold of the buffer is Vcc/2 and $V_N << Vcc$. Derive the time constant modulation based noise to jitter transfer function $d(td)/d(V_N)$ in terms of circuit parameter R, C, Vcc, and $dR/d(V_N)$.

[4.14] A simple RC based VLSI circuit jitter injection model is given in the figure below. Where the ideal buffer is assumed to have zero delay and parasitic capacitance. It has infinite gain and is clamped to 0 and Vcc.

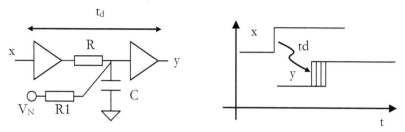

Assuming the threshold of the buffer is Vcc/2 and $V_N \ll V_{cc}$. Derive the cross-talk noise to jitter transfer function $d(td)/d(V_N)$ in terms of circuit parameter R, C, R1, and V_{CC}. Where R1 >> R.

[4.15] A simple RC based VLSI circuit jitter injection model is given in the figure below. Where the ideal buffer is assumed to have zero delay and parasitic capacitance. It has infinite gain and is clamped to 0 and Vcc.

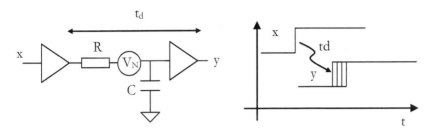

Assuming the threshold of the buffer is Vcc/2 and V_N<<Vcc. Derive the thermal noise induced jitter $\delta(td)$ in terms of circuit parameter R, C, T, and V_{CC}. Where thermal noise V_N can be approximately modeled as $(4kTR)^{0.5}$.

[4.16] A simple RC based VLSI circuit jitter injection model is given in the figure below. Where the ideal buffer is assumed to have zero delay and parasitic capacitance. It has infinite gain and is clamped to 0 and Vcc.

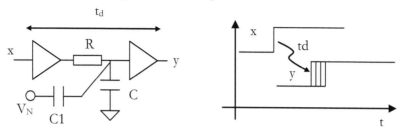

Assuming the threshold of the buffer is $Vcc/2$ and $V_N \ll Vcc$. Derive the cross-talk noise to jitter transfer function $d(td)/d(V_N)$ in terms of circuit parameter R, C, R1, and V_{CC}. Where $C \gg C1$.

[4.17] A simple RC based VLSI circuit jitter injection model is given in the figure below. Where the ideal buffer is assumed to have zero delay and parasitic capacitance. It has infinite gain and is clamped to 0 and Vcc.

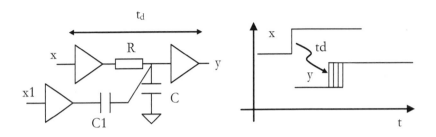

Assuming the threshold of the buffer is Vcc/2 and V_N<<Vcc. Also assuming x1 is random and synchronous to x. Derive the cross-talk noise induced jitter d(td) in terms of circuit parameter R, C, R1, and V_{CC}. Where C >> C1.

4.2 SAMPLE INTERVIEW QUESTIONS

1. What is the general temperature dependency of the circuit delay in CMOS technologies?

2. Explain why & how a MOS transistor works.

3. Draw a CMOS inverter. Explain its transfer characteristics.

4. Explain sizing selection methods of the CMOS inverter.

5. What is hot electron effect and how can it be eliminated?

6. Can both PMOS and NMOS transistors pass good 1 and good 0? Why and why not?

7. What is noise margin? What is the procedure to determine noise margin of a circuit?

8. How do you size NMOS and PMOS transistors in an inverter circuit to increase the threshold voltage?

9. What is electron migration effect and how it impacts the circuit delay? How can it be eliminated?

10. What is the effect of temperature on MOS device threshold voltage?

11. What is the effect of temperature on mobility?

12. What is the effect of gate voltage on mobility?

13. How can the rise and fall times in an inverter be equated?

14. If you have three adjacent parallel metal lines, two out of phase signals pass through the outer two metal lines. Draw the waveforms in the center metal line due to interference. Now, draw the signals if the signals in outer metal lines are in phase with each other.

15. How do you size NMOS and PMOS transistors to increase the threshold voltage?

16. Describe the various effects of VLSI scaling.

17. Draw a transistor level two input NAND gate. Explain its sizing (a) considering Vth (b) for equal rise and fall times.

18. What is the Miller Coefficient Factor (MCF)? How can it be used to model the delay effect of the VLSI interconnect circuits?

19. What does MCF = 0, 1, 1.5, 2 mean?

5

STATISTICAL VLSI CIRCUIT DELAY VARIATION MODELS

- Deterministic Jitter and Random Jitter
- Statistic Jitter Parameters
- Eye Diagram and Jitter Signature and Decomposition
- Total Jitter and Bathtub Curve Model

Skew and jitter are the static and dynamic delay deviations of the data or clock signals in high-speed I/O circuits. The PVT variations, the device mismatches, the voltage and phase noises, the reflection and ringing, and the channel bandwidth limitation effects cause these delay variations. Delay variations may significantly degrade the performances of the high-speed I/O circuits.

The macroscopic delay variation can usually be expressed using the eye-diagram of the signals. Statistical delay variation modeling methods can be used for identifying the sources of circuit delay variations in circuit debug, for predicting the circuit and system performance, such as the bit-error-rate (BER), for analyzing the jitter margin of the circuit and for predicting the yield of the implemented VLSI high-speed I/O circuits and systems at targeted data rates.

The jitter signature and jitter decomposition techniques based on the statistical delay variation analysis methods provide a very effective way for the post-silicon debug and the product validation in the practical high-speed I/O circuit design and development.

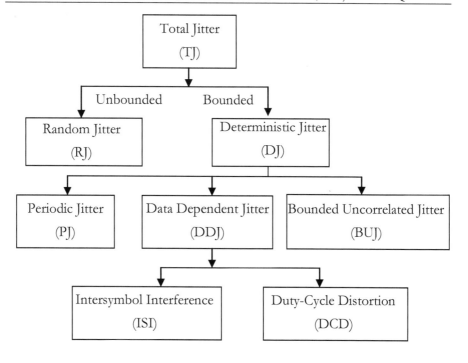

Fig.5.1 The jitter signatures and jitter decomposition

Bathtub curve provides a graphic method to express the relation between the eye opening of the signal within high-speed I/O with respect to the time of the measurement or BER in terms of the Q-factor.

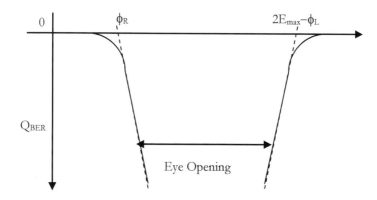

Fig.5.2 The bathtub curve

5.1 HOMEWORK AND PROJECT PROBLEMS

[5.1] A VLSI clock signal can be modeled as $V(t) = [1+\cos(\omega t + \phi_m \cos(\omega_m t))]/2$. Calculate the mean clock frequency, the peak-peak jitter, and the RMS jitter of the TIE. (Assuming $\omega = 2\pi$Ghz, $\phi_m = 0.1$, and $\omega_m = 2\pi*50$Mhz)

[5.2] For a VLSI clock signal modeled as $V(t) = [1+\cos(\omega t + \phi_m \cos(\omega_m t))]/2$. Calculate the peak-peak jitter, and the RMS jitter of the clock for period jitter. (Assuming $\omega = 2\pi$Ghz, $\phi_m = 0.1$, and $\omega_m = 2\pi*50$Mhz)

[5.3] For a VLSI clock signal modeled as $V(t) = [1+\cos(\omega t+\phi_m\cos(\omega_m t))]/2$. Calculate the peak-peak jitter, and the RMS jitter of the clock for cycle-to-cycle jitter. (Assuming $\omega = 2\pi$Ghz, $\phi_m = 0.1$, and $\omega_m = 2\pi*50$Mhz)

[5.4] For the above clock, generate the signature (PDF) of the TIE.

[5.5] For the above clock, generate the signature (PDF) of the period jitter.

[5.6] For the above clock, generate the signature (PDF) of the cycle-to-cycle jitter.

[5.7] A simple RC based VLSI circuit jitter injection model is given in the figure below. Where the ideal buffer is assumed to have zero delay and parasitic capacitance. It has infinite gain and is clamped to 0 and Vcc.

Assuming the threshold of the buffer is Vcc/2. Where C1 = C/20. Assume x is a 1Hz clock and D_N is a 1Hz square clock that is synchronized with x. Find the mean, RMS, and peak TIE for output clock y.

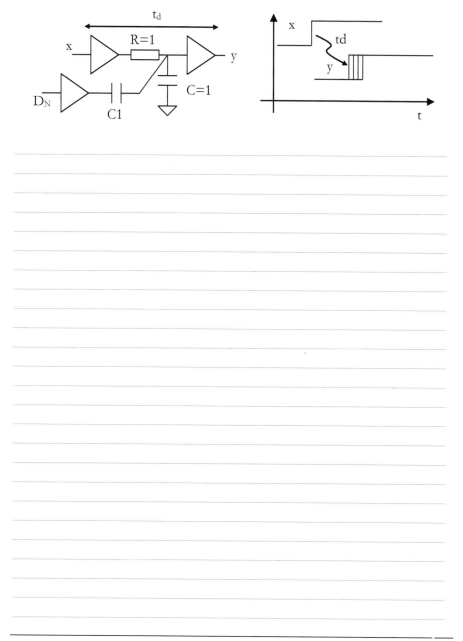

[5.8] A simple RC based VLSI circuit jitter injection model is given in the figure below. Where the ideal buffer is assumed to have zero delay and parasitic capacitance. It has infinite gain and is clamped to 0 and Vcc.

Assuming the threshold of the buffer is Vcc/2. Where $C1 = C/20$. Assume x is a 1Hz clock and D_N is a 1Hz square clock that is synchronized with x. Find the mean, RMS, and peak period jitter for output clock y.

[5.9] A simple RC based VLSI circuit jitter injection model is given in the figure below. Where the ideal buffer is assumed to have zero delay and parasitic capacitance. It has infinite gain and is clamped to 0 and Vcc.

Assuming the threshold of the buffer is Vcc/2. Where $C1 = C/20$. Assume x is a 1Hz clock and D_N is a 1Hz square clock that is synchronized with x. Find the mean, RMS, and peak cycle-to-cycle for output clock y.

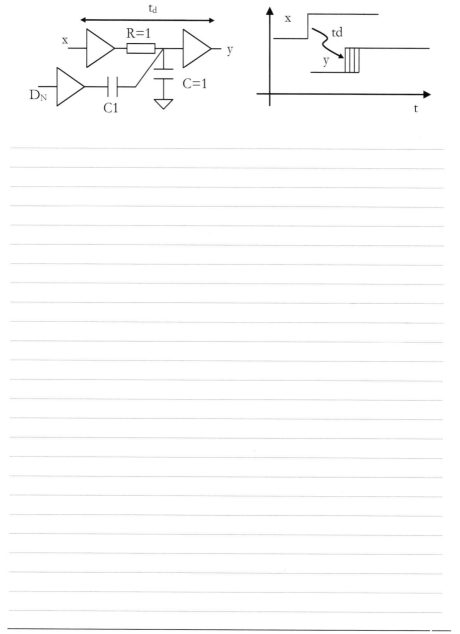

[5.10] A simple RC based VLSI circuit jitter injection model is given in the figure below. Where the ideal buffer is assumed to have zero delay and parasitic capacitance. It has infinite gain and is clamped to 0 and Vcc.

Assuming the threshold of the buffer is Vcc/2. Where C1 = C/20. Assume x is a 1Hz clock and D_N is a 1Hz square clock that is synchronized with x. Find the PDF of TIE for output clock y.

[5.11] A simple RC based VLSI circuit jitter injection model is given in the figure below. Where the ideal buffer is assumed to have zero delay and parasitic capacitance. It has infinite gain and is clamped to 0 and Vcc.

Assuming the threshold of the buffer is Vcc/2. Where C1 = C/20. Assume x is a 1Hz clock and D_N is a 1Hz square clock that is synchronized with x. Find the PDF for the period jitter for output clock y.

[5.12] A simple RC based VLSI circuit jitter injection model is given in the figure below. Where the ideal buffer is assumed to have zero delay and parasitic capacitance. It has infinite gain and is clamped to 0 and Vcc.

Assuming the threshold of the buffer is Vcc/2. Where C1 = C/20. Assume x is a 1Hz clock and D_N is a 1Hz square clock that is synchronized with x. Find the PDF for the cycle-to-cycle for output clock y.

[5.13] A simple RC based VLSI circuit jitter injection model is given in the figure below. Where the ideal buffer is assumed to have zero delay and parasitic capacitance. It has infinite gain and is clamped to 0 and Vcc.

Assuming the threshold of the buffer is Vcc/2. Where $C1 = C/20$. Assume x is a 1Hz clock and D_N is 0.5Hz clock that is synchronized with x. Find the mean, RMS, and peak TIE for output clock y.

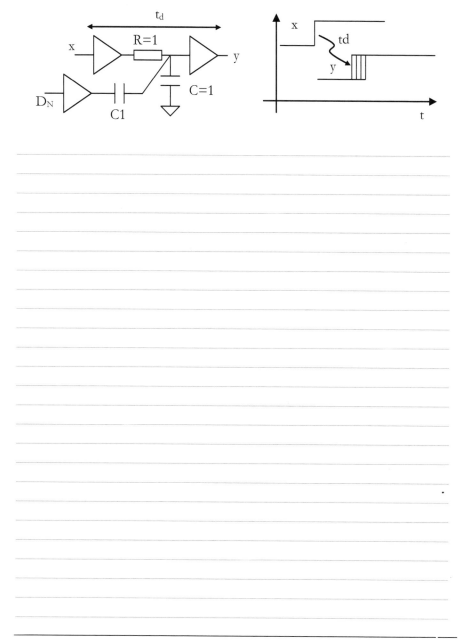

[5.14] A simple RC based VLSI circuit jitter injection model is given in the figure below. Where the ideal buffer is assumed to have zero delay and parasitic capacitance. It has infinite gain and is clamped to 0 and Vcc.

Assuming the threshold of the buffer is Vcc/2. Where C1 = C/20. Assume x is a 1Hz clock and D_N is 0.5Hz clock that is synchronized with x. Find the mean, RMS, and peak period jitter for output clock y.

[5.15] A simple RC based VLSI circuit jitter injection model is given in the figure below. Where the ideal buffer is assumed to have zero delay and parasitic capacitance. It has infinite gain and is clamped to 0 and Vcc.

Assuming the threshold of the buffer is Vcc/2. Where C1 = C/20. Assume x is a 1Hz clock and D_N is 0.5Hz clock that is synchronized with x. Find the mean, RMS, and peak cycle-to-cycle for output clock y.

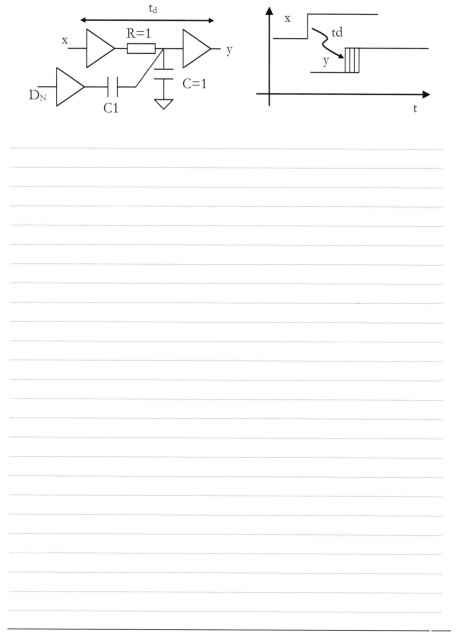

[5.16] A simple RC based VLSI circuit jitter injection model is given in the figure below. Where the ideal buffer is assumed to have zero delay and parasitic capacitance. It has infinite gain and is clamped to 0 and Vcc.

Assuming the threshold of the buffer is Vcc/2. Where C1 = C/20. Assume x is a 1Hz clock and D_N is 0.5Hz clock that is synchronized with x. Find the PDF of TIE for output clock y.

[5.17] A simple RC based VLSI circuit jitter injection model is given in the figure below. Where the ideal buffer is assumed to have zero delay and parasitic capacitance. It has infinite gain and is clamped to 0 and Vcc.

Assuming the threshold of the buffer is Vcc/2. Where C1 = C/20. Assume x is a 1Hz clock and D_N is 0.5Hz clock that is synchronized with x. Find the PDF for the period jitter for output clock y.

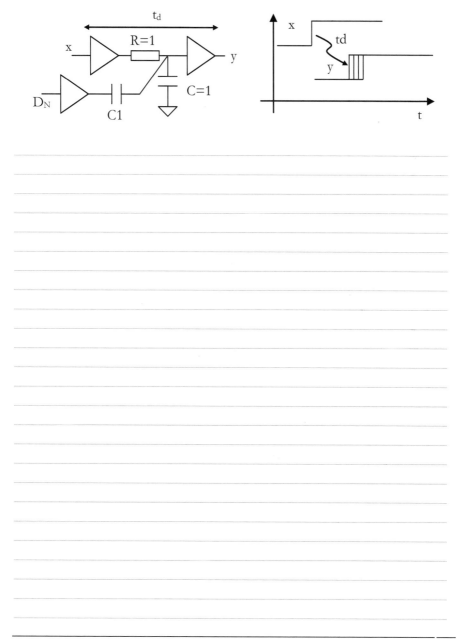

[5.18] A simple RC based VLSI circuit jitter injection model is given in the figure below. Where the ideal buffer is assumed to have zero delay and parasitic capacitance. It has infinite gain and is clamped to 0 and Vcc.

Assuming the threshold of the buffer is Vcc/2. Where C1 = C/20. Assume x is a 1Hz clock and D_N is 0.5Hz clock that is synchronized with x. Find the PDF for the cycle-to-cycle for output clock y.

[5.19] A simple RC based VLSI circuit jitter injection model is given in the figure below. Where the ideal buffer is assumed to have zero delay and parasitic capacitance. It has infinite gain and is clamped to 0 and Vcc.

Assuming the threshold of the buffer is Vcc/2. Where C1 = C/20. Assume x is a 1Hz clock and D_N is random clock. Find the mean, RMS, and peak TIE for output clock y.

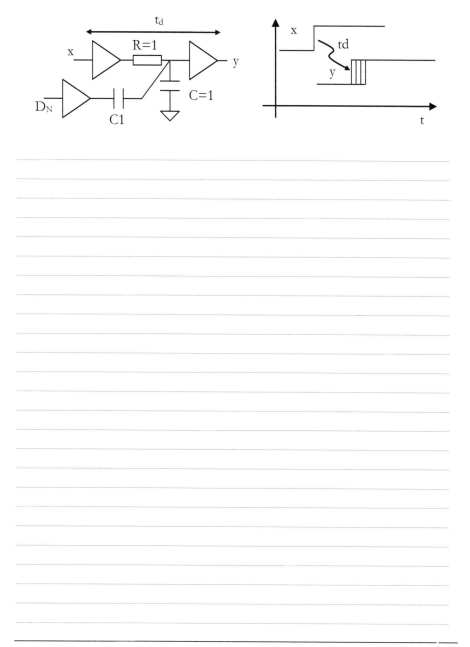

[5.20] A simple RC based VLSI circuit jitter injection model is given in the figure below. Where the ideal buffer is assumed to have zero delay and parasitic capacitance. It has infinite gain and is clamped to 0 and Vcc.

Assuming the threshold of the buffer is Vcc/2. Where $C1 = C/20$. Assume x is a 1Hz clock and D_N is random clock. Find the mean, RMS, and peak period jitter for output clock y.

[5.21] A simple RC based VLSI circuit jitter injection model is given in the figure below. Where the ideal buffer is assumed to have zero delay and parasitic capacitance. It has infinite gain and is clamped to 0 and Vcc.

Assuming the threshold of the buffer is Vcc/2. Where C1 = C/20. Assume x is a 1Hz clock and D_N is random clock. Find the mean, RMS, and peak cycle-to-cycle for output clock y.

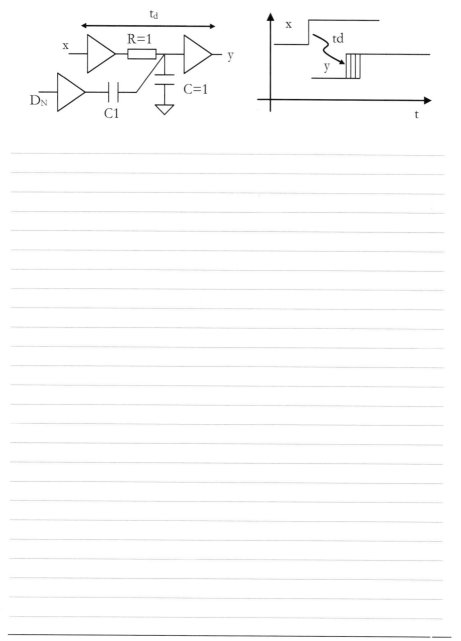

[5.22] A simple RC based VLSI circuit jitter injection model is given in the figure below. Where the ideal buffer is assumed to have zero delay and parasitic capacitance. It has infinite gain and is clamped to 0 and Vcc.

Assuming the threshold of the buffer is Vcc/2. Where C1 = C/20. Assume x is a 1Hz clock and D_N is random clock. Find the PDF of TIE for output clock y.

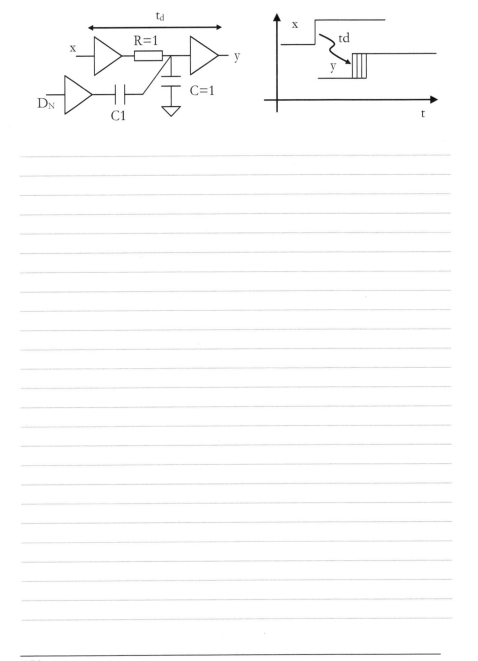

[5.23] A simple RC based VLSI circuit jitter injection model is given in the figure below. Where the ideal buffer is assumed to have zero delay and parasitic capacitance. It has infinite gain and is clamped to 0 and Vcc.

Assuming the threshold of the buffer is Vcc/2. Where $C1 = C/20$. Assume x is a 1Hz clock and D_N is random clock. Find the PDF for the period jitter for output clock y.

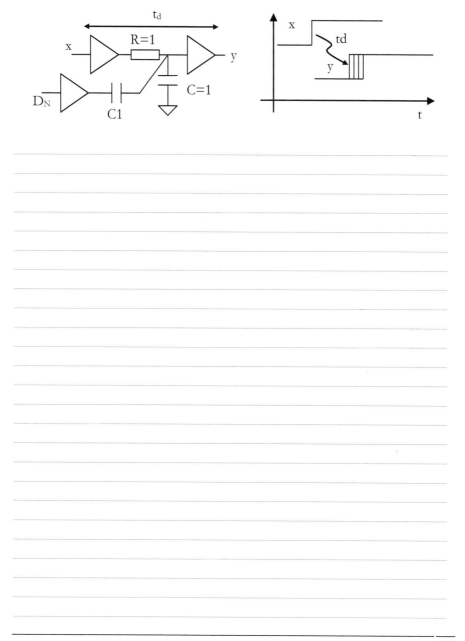

[5.24] A simple RC based VLSI circuit jitter injection model is given in the figure below. Where the ideal buffer is assumed to have zero delay and parasitic capacitance. It has infinite gain and is clamped to 0 and Vcc.

Assuming the threshold of the buffer is Vcc/2. Where C1 = C/20. Assume x is a 1Hz clock and D_N is random clock. Find the PDF for the cycle-to-cycle for output clock y.

[5.25] For a VLSI high-speed I/O circuit, if the TJ for BER = 10^{-10} and BER = 10^{-12} are 100ps and 140ps respectively, find the DJ and RJ (in terms of RMS) of the circuit.

[5.26] The bathtub curve of a VLSI high-speed I/O is given in figure below. Identify the regions of the curve in terms of Q-factor where DJ and RJ dominate respectively.

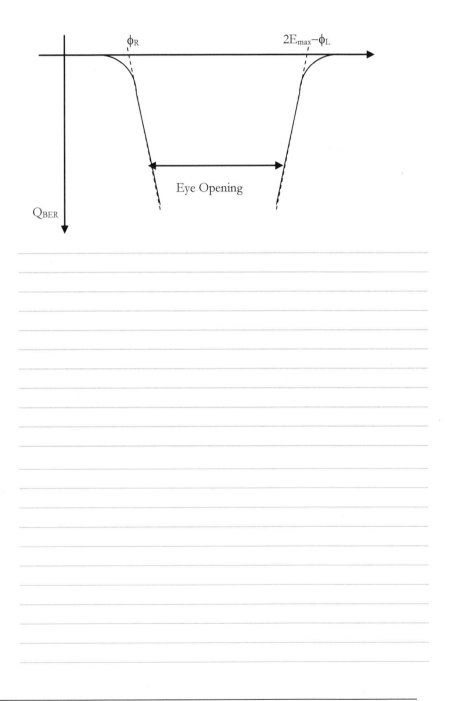

[5.27] In the dual Dirac jitter model, the slopes and intercepts of the two curves have specific meaning in the jitter specification. What jitter parameters are related to the two slopes and intercepts?

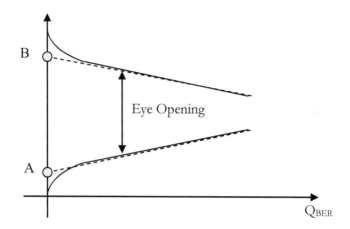

5.2 SAMPLE INTERVIEW QUESTIONS

1. Why RJ is usually not specified using it peak-to-peak value in VLSI high-speed I/O circuit?

2. What is a jitter and how many types of jitters do you know?

3. What is a typical VLSI high-speed I/O Rx jitter tolerance curve?

4. Will a high-speed I/O circuit tolerance be larger than 1 UI jitter?

5. How jitter can be applied to a design?

6. What is jitter? Explain how occurs? How to overcome jitter?

7. What is skew? How does it impact performance of the high-speed I/O circuits?

8. What is the bathtub curve?

9. What is the Q-factor in bathtub curve? What is the Q-factor for BER = 10E-12?

10. What is the general dependency of eye opening on the BER?

11. What is the general dependency of the eye opening on the DJ of the high-speed I/O circuit?

12. If there are two RJ sources in a high-speed I/O circuit and each contributes to 1ps RMS RJ. What is likely the total RJ of the circuit?

13. If there are two DJ sources in a high-speed I/O circuit and each one contributes to 10ps peak DJ, what is the best and worst peak DJ value of the circuit?

14. Please explain the dependency of the measured eye opening in the eye-diagram versus the time of the measurement.

6

VLSI DELAY TIME AND PHASE CONTROL CIRCUITS

- Voltage Controlled Delay Line
- Voltage Controlled Oscillator
- Phase Interpolator

Electrically controllable delay circuits are widely used in VSI high-speed I/O for compensating the circuit delay uncertainties. There are three major VLSI delay control circuit families, including the voltage control delay line (VCDL) circuits, the voltage control oscillator (VCO) circuits and the phase interpolator (PI) circuits.

VLSI voltage controlled delay line (VCDL) and voltage controlled oscillator (VCO) circuits offer time-domain signal processing capability, which allows electrically controllable delay time of the signals. The control signals of VCDL and VCO circuits are typically analog voltages signals. However, they can also be in other analog (such as current) or digital signals forms. VLSI phase interpolation (PI) circuits, on the other hand, usually serve as the digital to delay converters that are based on phase weighting of multiple phase clock inputs.

VLSI voltage controlled delay line, voltage controlled oscillator and phase interpolator circuits are usually used in the phase control loops in VLSI high-speed I/O circuits implemented in the forms of delay-locked loop (DLL) circuits, phase-locked loop (PLL) circuits and data recover circuits (DRC) for adaptive compensations of circuit or system delay time uncertainties.

6.1 HOMEWORK AND PROJECT PROBLEMS

[6.1] A simple RC based VLSI VCDL circuit is shown in the figure below. Where the ideal buffer is assumed to have zero delay and parasitic capacitance. It has infinite gain and is clamped to 0 and Vcc.

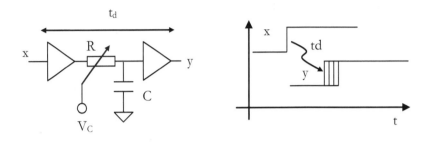

Assuming the threshold of the buffer is Vcc/2. Find the gain of the VCDL in terms of R, C, and dR/d(V$_C$).

[6.2] A simple RC based VLSI VCDL circuit is shown in the figure below. Where the ideal buffer is assumed to have zero delay and parasitic capacitance. It has infinite gain and is clamped to 0 and Vcc.

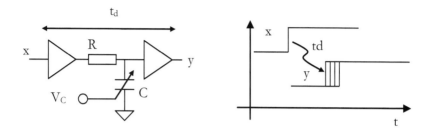

Assuming the threshold of the buffer is Vcc/2. Find the gain of the VCDL in terms of R, C, and dC/d(V$_C$).

[6.3] A simple RC based VLSI VCDL circuit is shown in the figure below. Where the ideal buffer is assumed to have zero delay and parasitic capacitance. It has infinite gain and is clamped to 0 and Vcc.

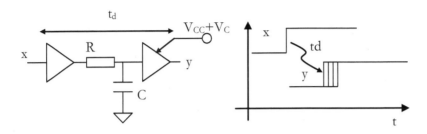

Assuming the threshold of the buffer is Vcc/2. Find the gain of the VCDL in terms of R, C, and V_{CC}.

[6.4] Shown in figures are the simulated VCDL delay time versus the control voltage. Determine the ϕ_i and K_{VCO} of this circuit. T is the period of the reference clock.

[6.5] A VLSI VCO is constructed using identical inverted VCDL elements with delay versus control voltage given in figure below. Calculate the frequency tuning range for this VCO with 0.2<Vcnt<0.8. Determine the VCO gain for the circuit.

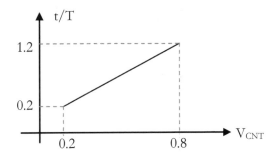

[6.6] Two commonly used VLSI VCDL circuit structures are shown in figure below. Explain how they work and the major differences between the two circuit architectures.

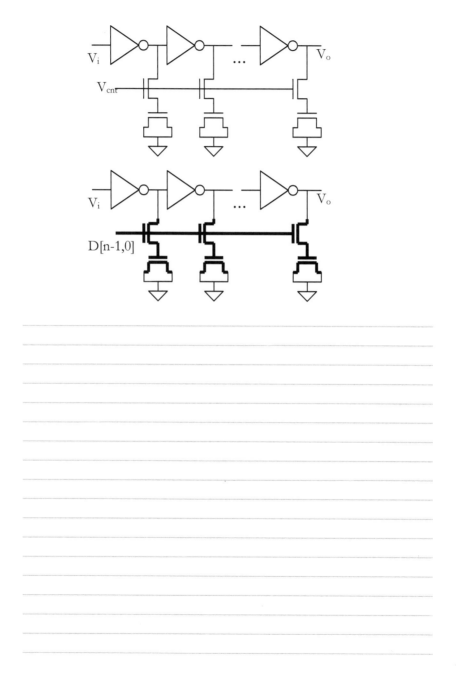

[6.7] Shown in figure below is a very popular VLSI VCDL circuit structure. Explain the operation of the circuit. How will the voltage swing of the VCDL circuit depend on the VCDL control voltage?

Vcnt

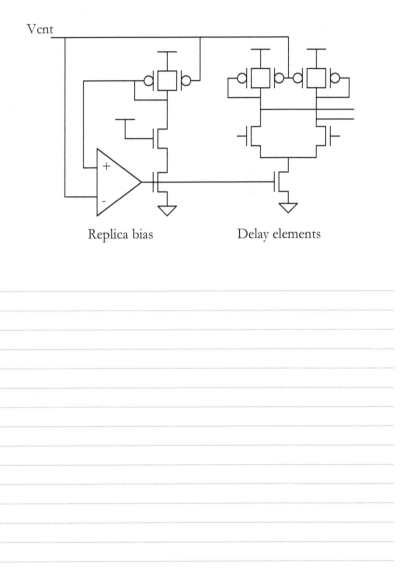

Replica bias   Delay elements

[6.8] A VLSI LC VCO and its circuit model are shown in figure below. Derive the VCO self-oscillation condition and the oscillation frequency in terms of the given circuit parameters.

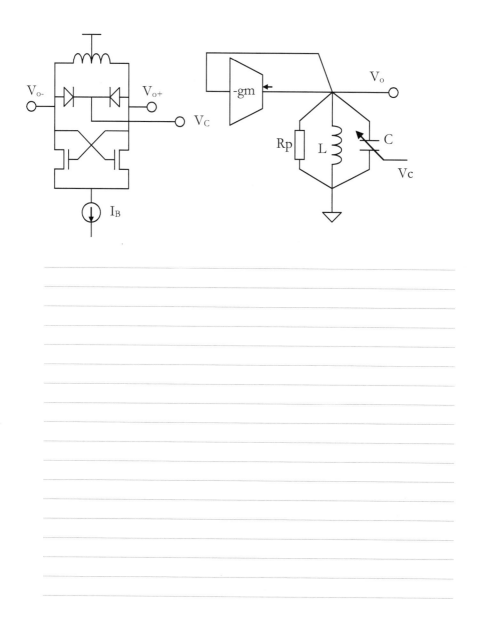

[6.9] For the sinusoidal phase interpolation algorithm shown below, find the coefficient α_I and α_Q such that $\phi = \pi/4$.

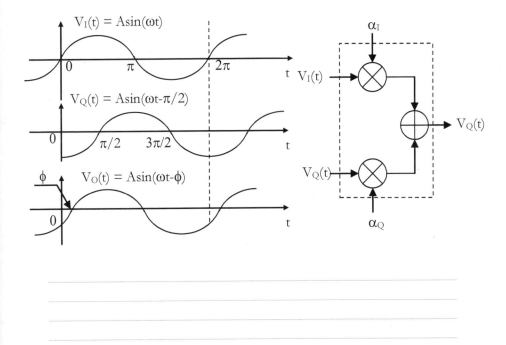

[6.10] Prove your result in [6.9] using circuit simulation.

[6.11] Design a VLSI PI using mixer circuit shown in figure below which uses a 0.18um CMOS technology. Simulation I1 and I2 for the 0°, 22.5°, 45o, 67.5°, 90° output clock phases. Assuming 1Ghz sinusoidal input clock signals. Plot the phase diagram of this PI. Please include as CML to CMOS buffer in your design.

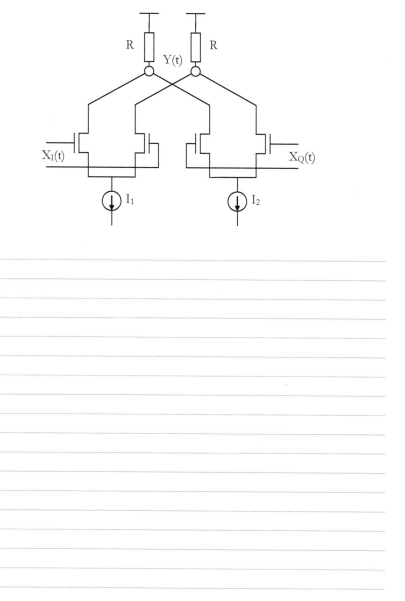

[6.12] Design a VLSI PI using mixer circuit shown in figure below which uses a 0.18um CMOS technology. Simulation R1 and R2 for the 0°, 22.5°, 45o, 67.5°, 90° output clock phases. Assuming 1Ghz square wave input clock signals. Plot the phase diagram of this PI. Determine your optimal C value in your design.

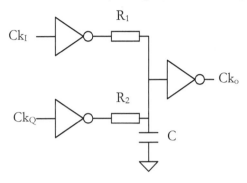

[6.13] Design a VLSI 20-to-1 PISO circuit that works at 1Ghz clock using the 0.18um CMOS technology.

[6.14] Design a VLSI 1-to-20 SIPO circuit that works at 1Ghz clock using the 0.18um CMOS technology

6.2 SAMPLE INTERVIEW QUESTIONS

1. What is a voltage controlled delay line (VCDL)?

2. What is a voltage-controlled oscillator (VCO)?

3. What is a phase interpolator (PI)?

4. How is the VCDL modeled? What is the relation between the VCDL output phase and the control voltage?

5. How is the VCO modeled? What is the relation between the VCO output phase and the control voltage?

6. How is the PI modeled? What is the relation between the PI output phase and the control codes?

7. What are the two commonly used PI phase interpolation algorithms? What are the key advantages and disadvantages of the two approaches?

8. Describe the key circuit blocks in a VLSI PI circuit.

9. What is the PI linearity? What circuit parameters may impact VLSI PI linearity?

10. List the VLIS PI circuit implementation you know.

11. What typically does a resistor division PI belong to? What is the major linearity cause of the voltage-mode PI circuits?

7

VLSI SYNCHRONIZATION CIRCUITS

- Bistable Circuit Structure
- VLSI Latch and Flip Flop Circuits
- SIPO
- PISO

Synchronization is a fundamental VLSI time-domain signal processing operation in where the delay (phase) of a data signal is "forced" to a known time reference. VLSI synchronization circuits are basic circuit elements in VLSI high-speed I/O circuits for data signal phase noise filtering and for received data sampling. Synchronization can be realized in various VLSI circuits such as static latch (or Flip Flop), dynamic latch or mux circuits. Static latch synchronization circuits are the most commonly used VLSI synchronization circuits. VLSI dynamic latch and multiplexes based VLSI synchronization circuits are increasingly used in high data rate VLSI high-speed I/O circuit applications.

The performance of VLSI synchronization circuits can usually be expressed using the sampler performance parameters, such as the sampler setup and hold times, the clock to output delay time, the transparent datapath delay, and the input voltage thresholds.

VLSI synchronization circuits employing the basic bi-stable circuits need to meet the meta-stability constraints that set the setup hold times, voltage thresholds and the clock to output delay parameters of the circuits.

VLSI synchronization circuits are - used for high-speed I/O data transmission synchronization and for data sampling. They are also used for the cross clock domain synchronization, frequency division, phase detection, and data rate conversions, such as parallel in serial out (PISO) and serial in parallel out (SIPO) circuit operations.

7.1 HOMEWORK AND PROJECT PROBLEMS

[7.1] For the VLSI sampler circuit shown in figure below, assuming both buffers are ideal that has zero input parasitic capacitance, zero output resistance, zero delay and infinite gain with $V_{CC}/2$ threshold. Assuming the pass-gate switch has an equivalent NO-resistance of R. Determine V_{IL}, V_{IH}, as well as the worst case setup time, hold time and TCO of circuit in terms of R and C values (assuming C is fully charged or discharged at the initial condition).

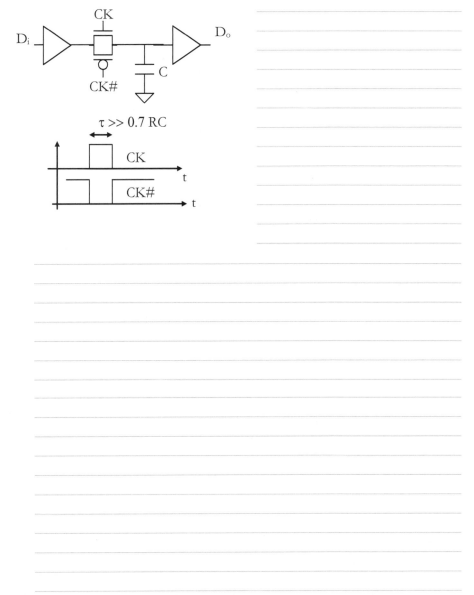

[7.2] Shown in figure below is a simple VLSI sampler circuit. Assuming both buffers are ideal that has zero input parasitic capacitance, zero output resistance, zero delay and infinite gain with $V_{CC}/2$ threshold. Assuming the pass-gate switch has an equivalent ON-resistance of R. Determine V_{IL}, V_{IH}, as well as the worst case setup time, hold time and TCO of circuit in terms of R and C values (assuming C is fully charged or discharged at the initial condition).

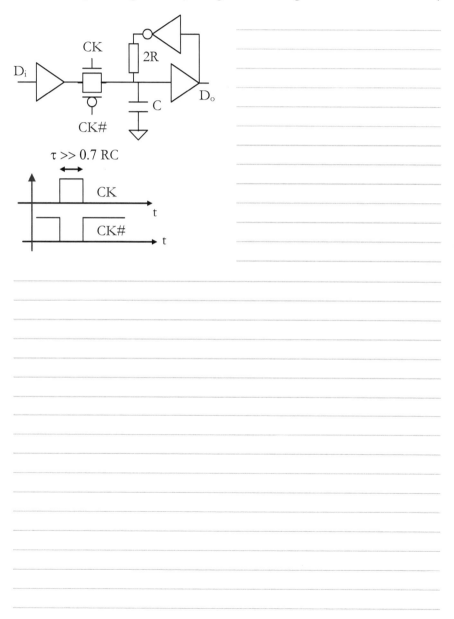

[7.3] Shown in figure below is a simple VLSI sampler circuit. Assuming both buffers are ideal that has zero input parasitic capacitance, zero output resistance, zero delay and infinite gain with $V_{CC}/2$ threshold. Assuming the pass-gate switch has an equivalent ON-resistance of R. Determine V_{IL}, V_{IH}, as well as the worst case setup time, hold time and TCO of circuit in terms of R and C values (assuming C is fully charged or discharged at the initial condition).

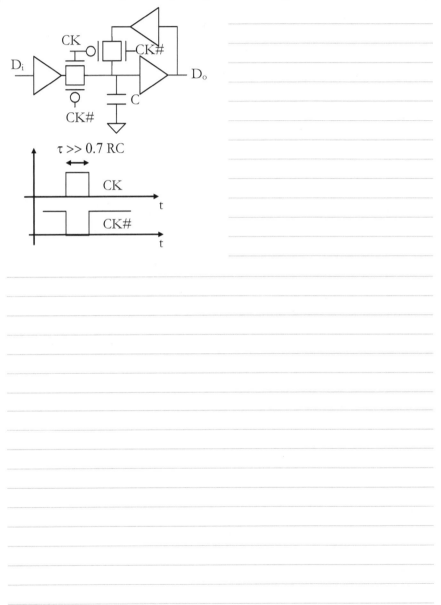

[7.4] For problem [7.1], [7.2], and [7.3] how the results will be different if we get rid of the input buffer?

[7.5] The transfer function of a VLSI bistable circuit structure is shown in below. Determine the state transition path and the final state of the circuit if the initial state of the circuit is given as point S

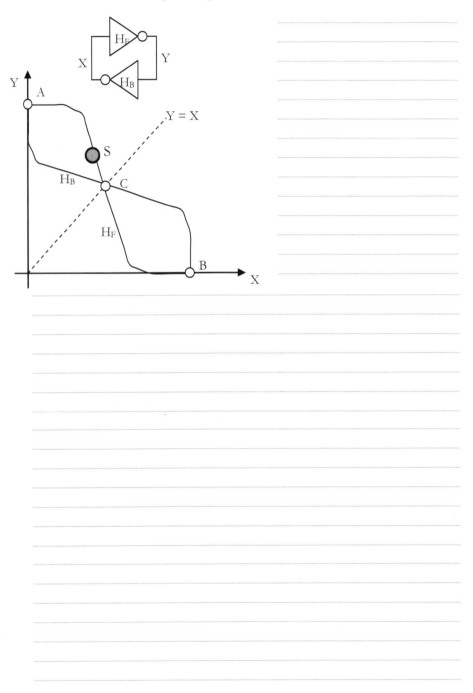

[7.6] The transfer function of a VLSI bistable circuit structure is shown in below. Determine the state transition path and the final state of the circuit if the initial state of the circuit is given as point S.

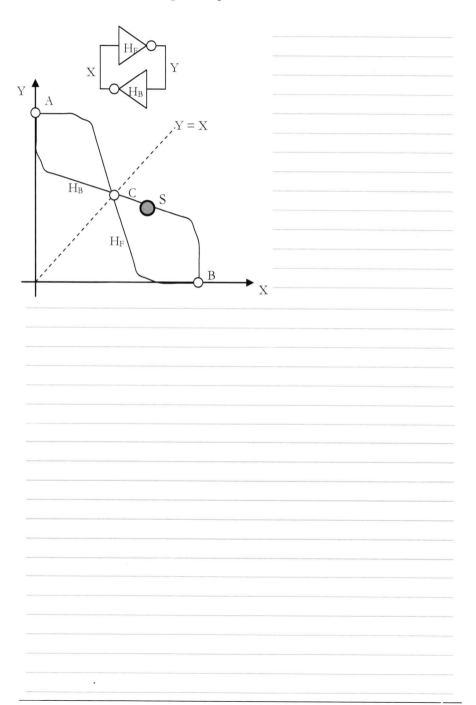

[7.7] The transfer function of a VLSI bistable circuit structure is shown in below. Determine the state transition path and the final state of the circuit if the initial state of the circuit is given as point S.

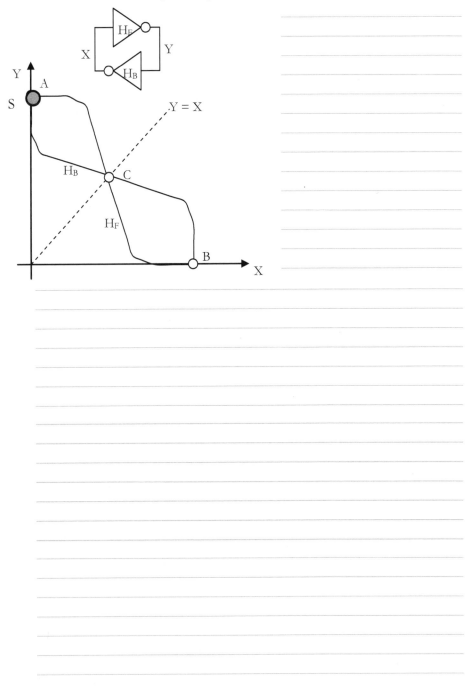

[7.8] The transfer function of a VLSI bistable circuit structure is shown in below. Determine the state transition path and the final state of the circuit if the initial state of the circuit is given as point S.

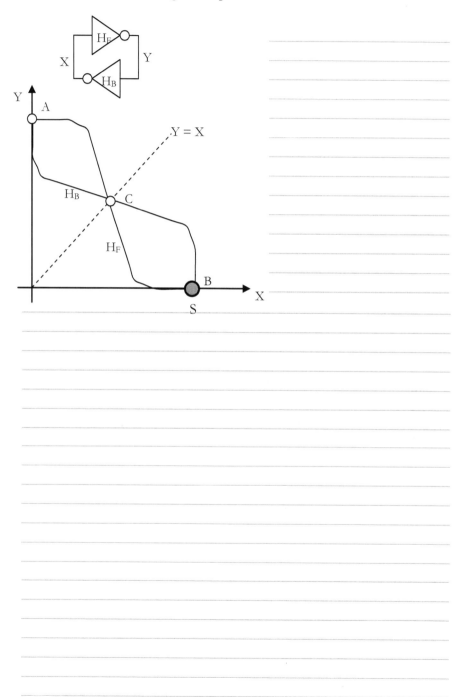

[7.9] The transfer function of a VLSI bistable circuit structure is shown in below. Determine the state transition path and the final state of the circuit if the initial state of the circuit is given as point S.

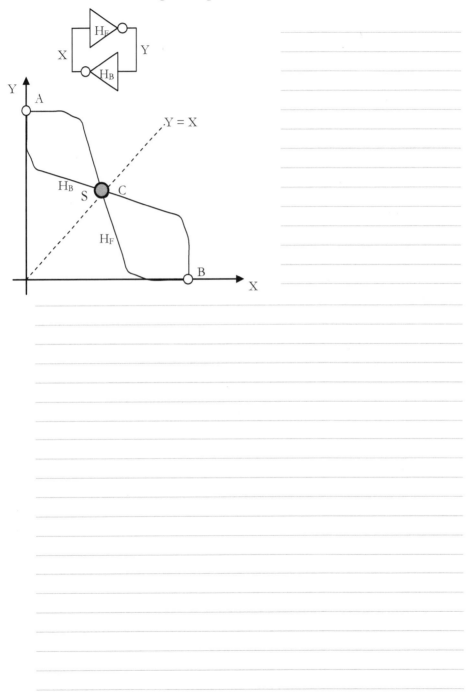

[7.10] If the transfer function of the basic VLSI bistable circuit structure (R = infinite) is given in the figure below. What are likely the new transfer function curves for finite R values?

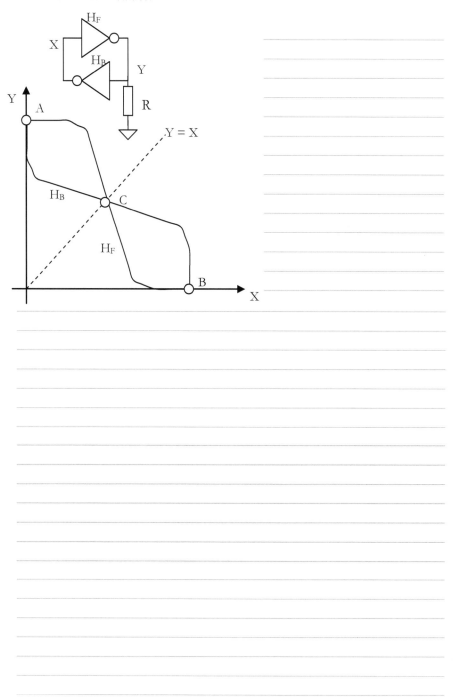

[7.11] If the transfer function of the basic VLSI bistable circuit structure (R = infinite) is given in the figure below. What are likely the new transfer function curves for finite R values?

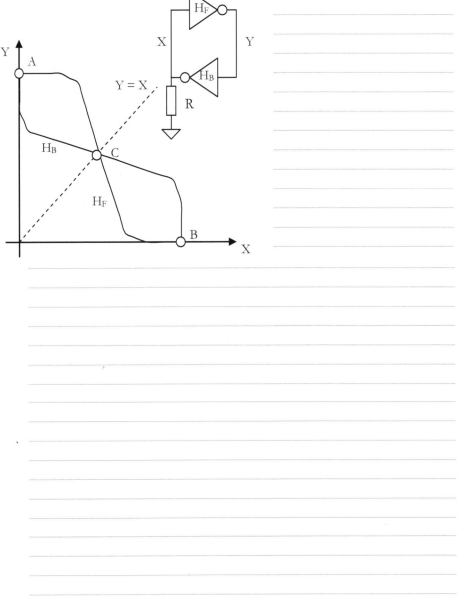

[7.12] If the transfer function of the basic VLSI bistable circuit structure (R = infinite) is given in the figure below. What are likely the new transfer function curves for finite R values?

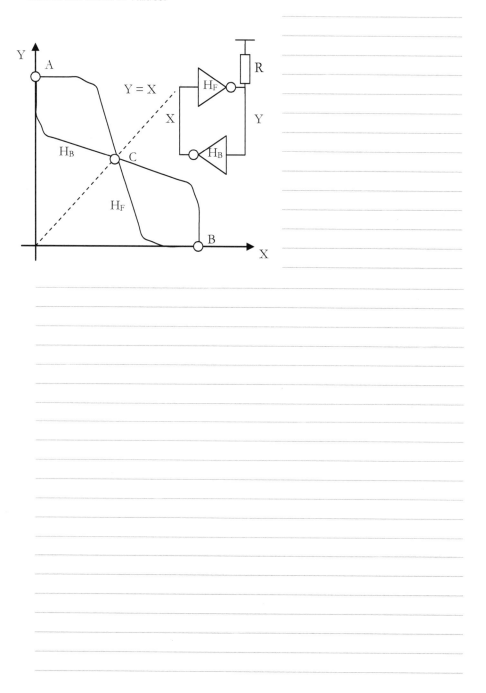

[7.13] If the transfer function of the basic VLSI bistable circuit structure (R = infinite) is given in the figure below. What are likely the new transfer function curves for finite R values?

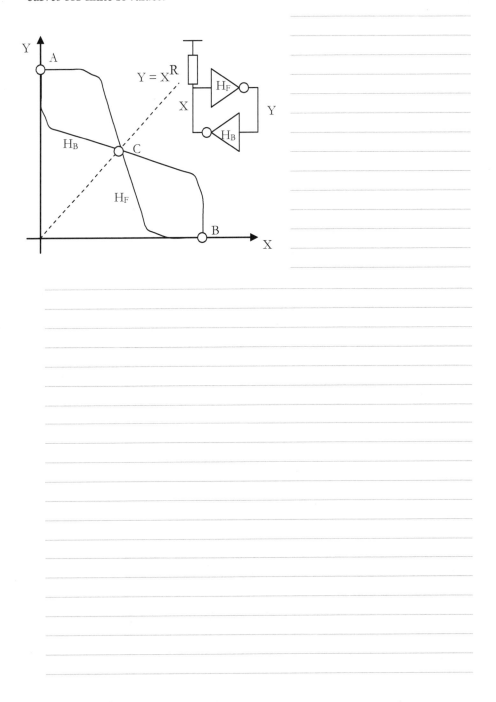

[7.14] If the transfer function of the basic VLSI bistable circuit structure is given in the figure below. What are likely the new transfer function curves if the width of the PMOS device in H_F is increased by 2x?

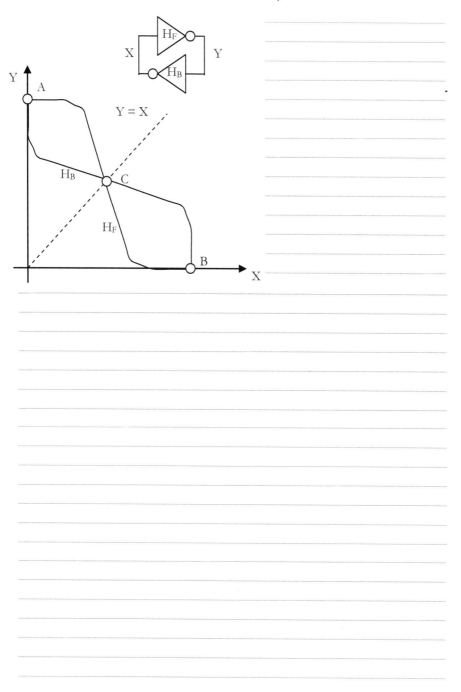

[7.15] If the transfer function of the basic VLSI bistable circuit structure is given in the figure below. What are likely the new transfer function curves if the width of the PMOS device in H_F is reduced by half?

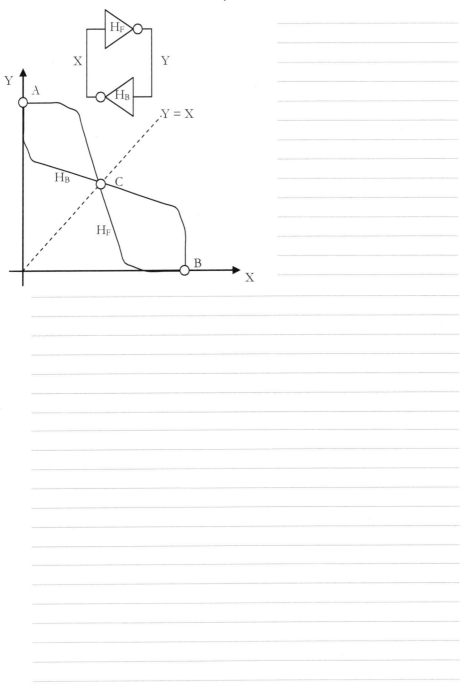

[7.16] If the transfer function of the basic VLSI bistable circuit structure is given in the figure below. What are likely the new transfer function curves if the width of both PMOS and NMOS devices in H_F is increased by 2x?

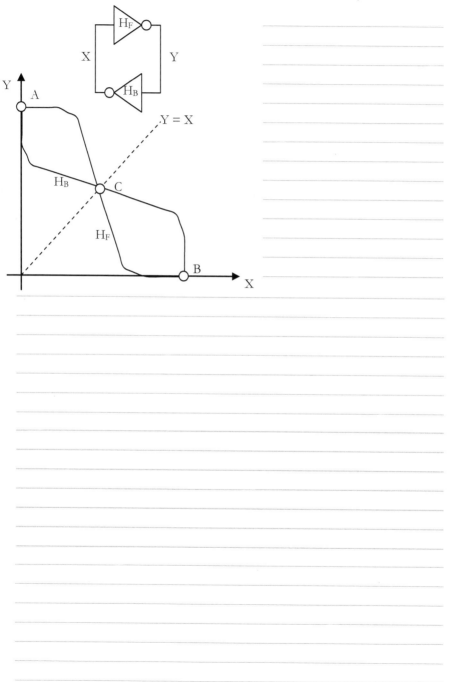

[7.17] If the transfer function of the basic (E = 0) VLSI bistable circuit structure is given in the figure below. What are likely the new transfer function curves for finite E values?

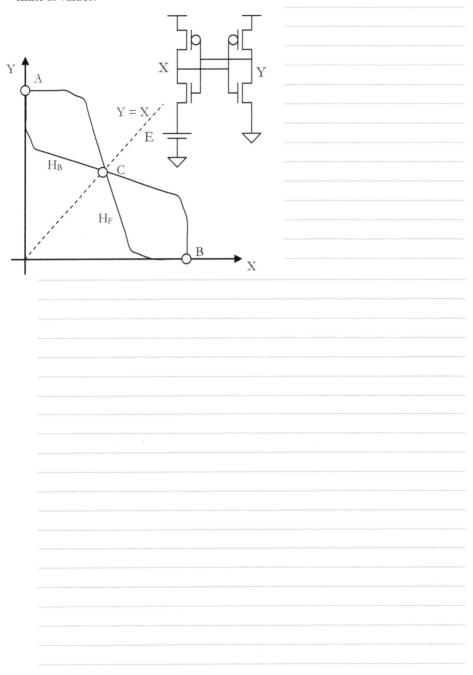

[7.18] If the transfer function of the basic ($E_X = E_Y = 0$) VLSI bistable circuit structure is given in the figure below. What is likely the new transfer function curves for finite values and $E_X > E_Y$?

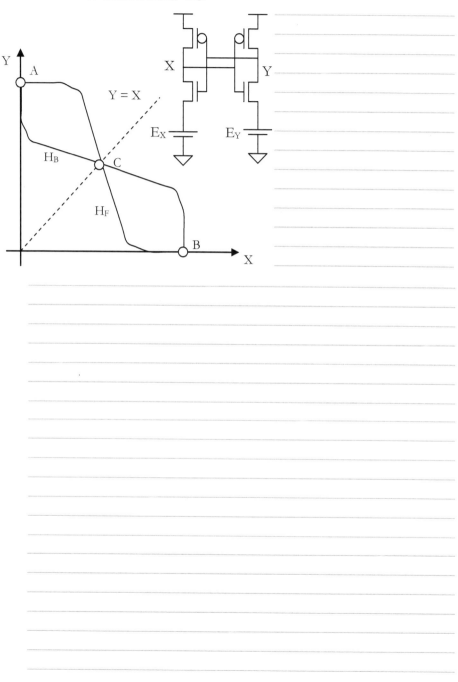

[7.19] If the transfer function of the basic (S $=$ R $=$ 1) VLSI bistable circuit structure is given in the figure below. What is likely the new transfer function curves for S $=$1 and R $=$ 0?

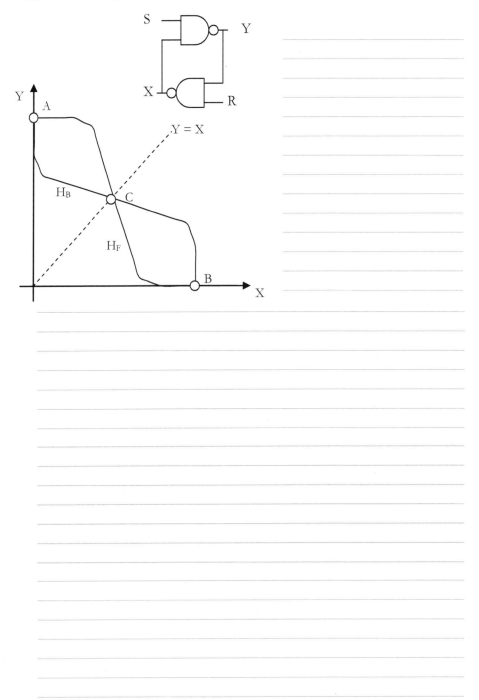

[7.20] If the transfer function of the basic (S =R = 1) VLSI bistable circuit structure is given in the figure below. What is likely the new transfer function curves for S =0 and R = 1?

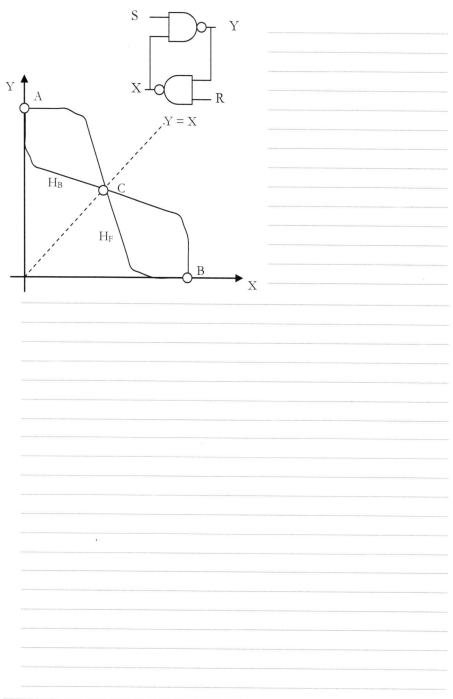

[7.20] If the transfer function of the basic (S =R = 1) VLSI bistable circuit structure is given in the figure below. What is likely the new transfer function curves for S = 0 and R = 0?

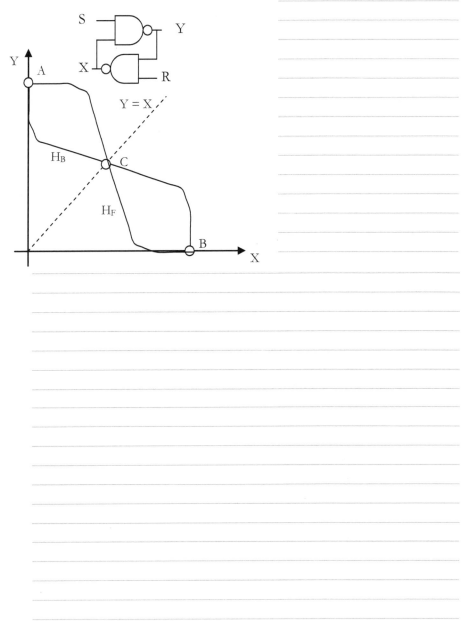

[7.22] If the transfer function of the basic VLSI bistable circuit structure is given in the figure below. What is likely the new transfer function curves if capacitors Cx and Cy are added to the circuit?

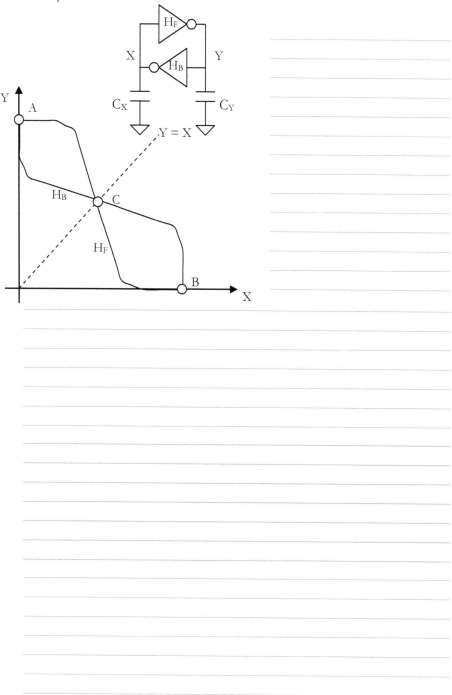

[7.23] If the basic VLSI bistable circuit structure is modeled as figure below. Estimate the transition time for the circuit to transfer from the initial state to the final stable state. (Assuming both inverters is identical. Ignoring the parasitic capacitance of the inverters. Assuming $V_X = V_{CM} + V_I$ and $V_Y = V_{CM} - V_I$), where V_{CM} is the node voltage at point C)

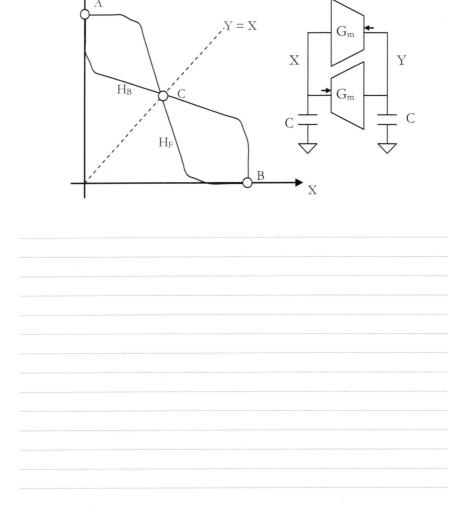

[7.24] Simulate for the setup time, hold time, and the CTO for the following dynamic latch using a 0.18um CMOS technology. (for simplicity, use 0.24/0.18 for all devices and let FO = 3).

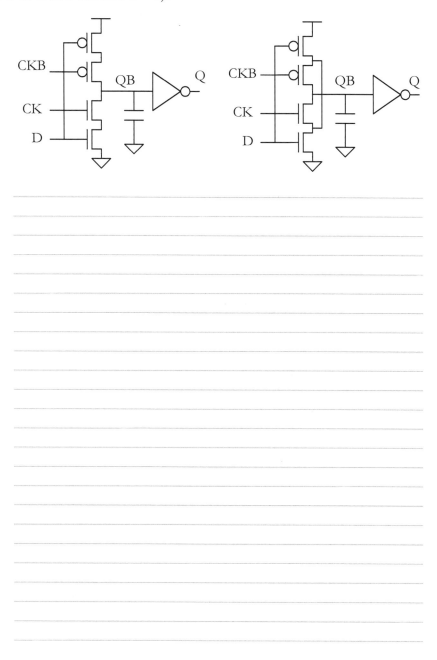

[7.25] Simulate for the setup time, hold time, and the CTO for the following dynamic latch using a 0.18um CMOS technology. (for simplicity, use 0.24/0.18 for all devices and let FO = 3).

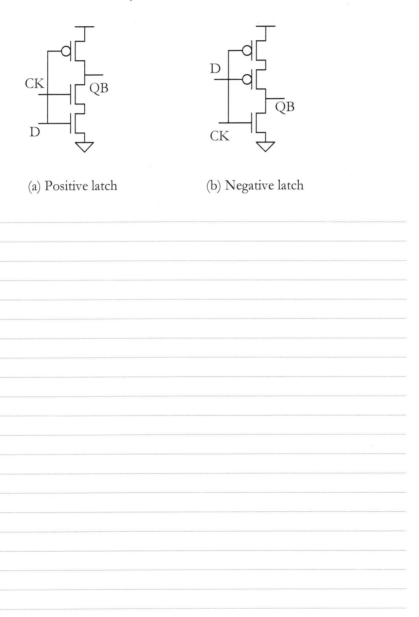

(a) Positive latch   (b) Negative latch

[7.26] Simulate for the setup time, hold time, and the CTO for the following dynamic flip-flop using a 0.18um CMOS technology. (for simplicity, use 0.24/0.18 for all devices and let FO = 3).

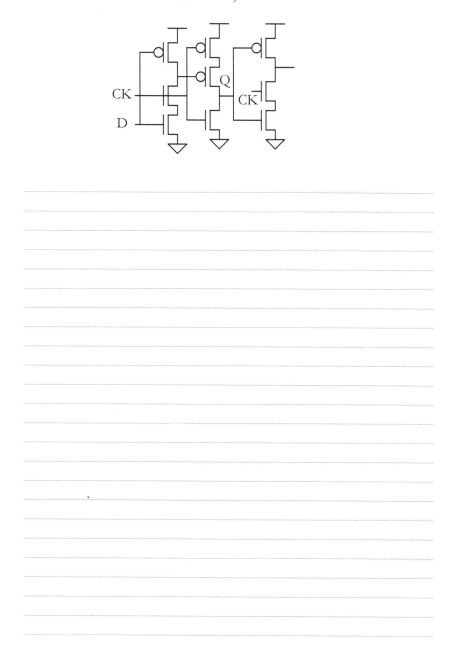

[7.27]. Describe how the following MUX multi-path synchronization works.

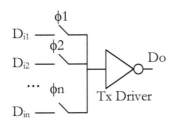

[7.28]. Simulate for the setup time, hold time, and the CTO for above MUX multi-path synchronization circuit using a 0.18um CMOS technology. (for simplicity, use 0.24/0.18 for all devices and let FO = 3).

[7.29] Design a 1Gbps PISO circuit based on the structure shown below using a 0.18um CMOS technology.

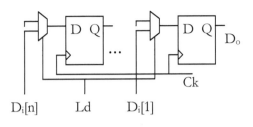

[7.30] Design a 1Gbps PISO circuit based on the structure shown below using 0.18um CMOS technology.

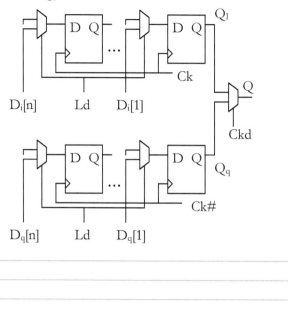

[7.31] Design a 1Gbps Tx synchronization circuit based on the structure shown below using 0.18um CMOS technology.

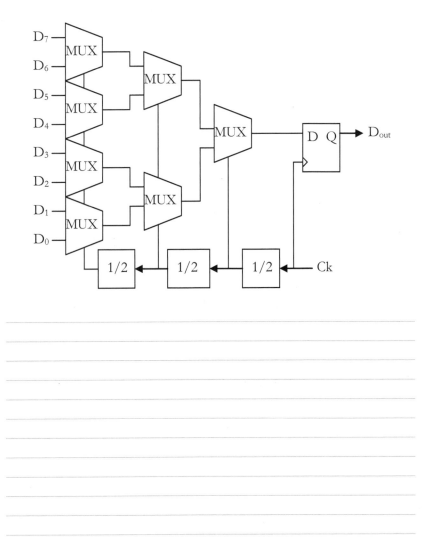

7.2 SAMPLE INTERVIEW QUESTIONS

1. Why PMOS and NMOS are usually sized equally in a transmission gates?

2. What is metastability? When/why it will occur? What are the different ways to avoid this?

3. What is meant by single phase and double phase clocking?

4. What is the difference between flip-flop and latch?

5. What are the critical parameters in a latch and flip-flop?

6. How did you arrive at sizes of a 6-transistor keeper cell structure that commonly used in latch?

7. What are different types of flip-flops?

8. Implement D- latch from (a) RS flip flop; (b) multiplexer.

9. How to convert D-latch into JK-latch and JK-latch into D-latch?

10. You have two 4-bit counters, built from D-FF. First circuit is synchronous and second is "ripple" (cascading). Which circuit has a less propagation delay?

11. Draw a 6-transistor keeper circuit and explain its operations

12. Draw the timing diagram for a 6-T keeper. What happens if we delay the enabling of clock signal?

13. Explain how a Flip flop works. What is metastable state in flip-flops?

14. Construct a D-FF from a T-FF.

15. Draw the state graphs for a given problem like sequence generator, flip flop.

16. How many 4:1 mux do you need to design an 8:1 mux?

17. What is a D-latch? Implement D flip-flop with a couple of latches.

18. How can you convert an SR Flip-flop to a JK Flip-flop?

19. How can you convert a JK Flip-flop to a D Flip-flop?

20. What is race-around problem? How can you rectify it?

21. What is the difference between synchronous design and asynchronous design?

22. How was clock-crossing accomplished? What are synchronizers, metastability, and determinism?

23. For a FIFO with input and output data using a 100Mhz and a 80Mhz clocks respectively, there are only 80 data input in any order during each 100 clocks. In other words, a 100 input clock will carry only 80 data and the other twenty clocks carry no data (data is scattered in any order). How big the FIFO needs to be to avoid data over/under flow.

24. How to synchronize control signals and data between two different clock domains?

25. What is the difference between latches and flip-flops based designs

8

VLSI DELAY AND PHASE COMPUTATIONAL CIRCUITS

- Basic VLSI Delay Computations
- VLSI Phase Scaling Operations

Basic time-domain signal processing operations such as the addition, the scaling and the integration operations can be realized using VLSI circuits. These basic VLSI time-domain signal processing operations play important roles in adaptive delay time and delay variation cancellation and in control loop compensation for VLSI high-speed I/O circuits.

Delay addition can be directly realized by connecting multiple VLSI delay circuit elements in series. Delay scaling operation can be implemented based on replica delay circuit elements or delay interpolation circuit elements. Time-domain integration operation, on the other hand, can either be realized employing VCO circuits or employing VLSI signal integration circuit and VLSI voltage to time conversion circuit elements, where VCO circuit provides almost perfect time-domain integration operation.

Phase scaling operations are usually realized through various frequency division circuit techniques. Frequency division provides near perfect scaling of the delay phase of a time-domain signal through the scaling of the reference clock period. VLSI frequency dividers are commonly used in the VLSI PLL circuits for reference clock generation operations.

8.1 HOMEWORK AND PROJECT PROBLEMS

[8.1] Determine the division factor for the VLSI frequency divider circuit shown below. Show the signal diagrams of the key nodes.

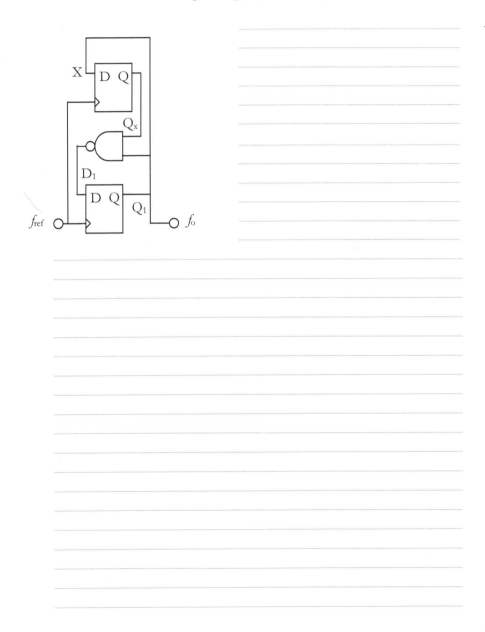

[8.2] Determine the division factor for the VLSI frequency divider circuit shown
below. Show the signal diagrams of the key nodes.

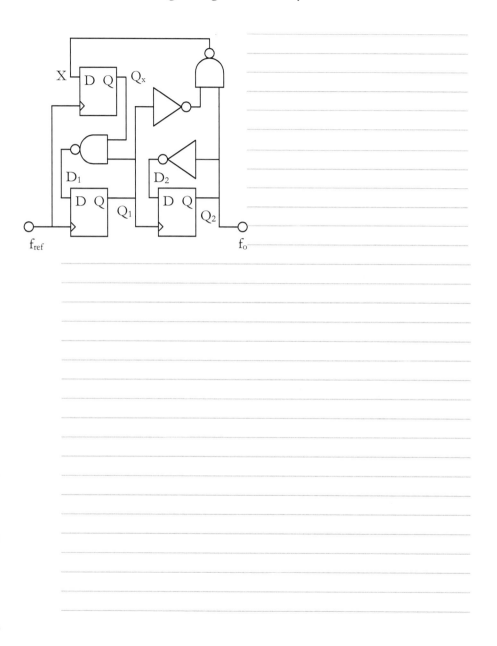

[8.3] Prove the following delay element combinations are equivalent.

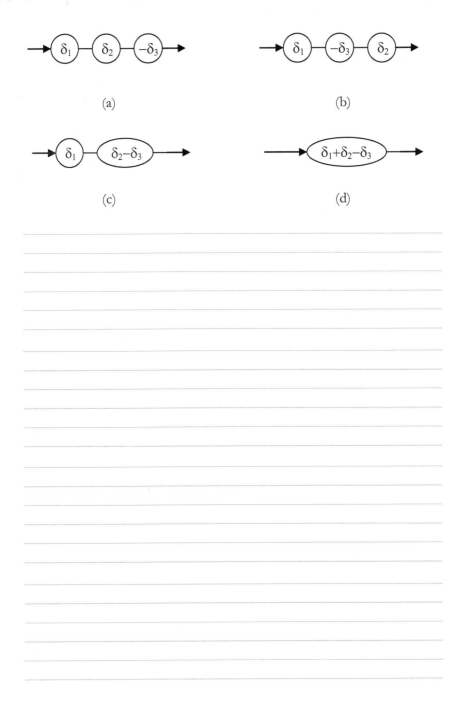

(a)

(b)

(c)

(d)

[8.4] Prove the two circuits are equivalent.

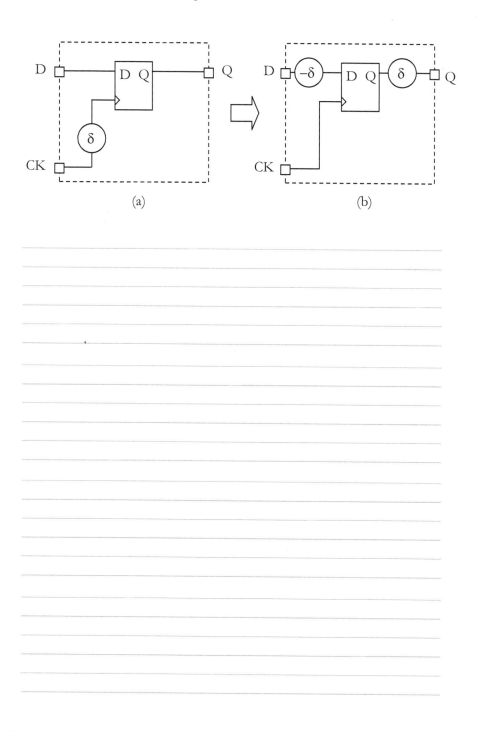

(a) (b)

[8.5] Prove the two circuits are equivalent.

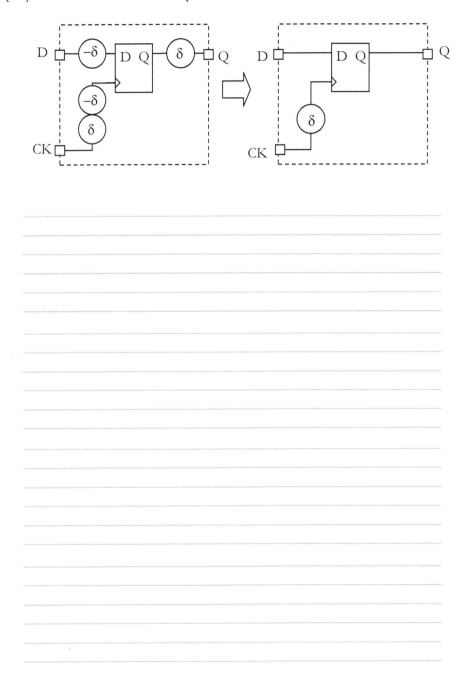

[8.6] Design a divide by 2 circuit.

[8.7] Design a divide-by-3 sequential circuit with 50% duty circle.

[8.8] Design a finite state machine to give a modulo 3 counter when x=0 and modulo 4 counter when x=1.

[8.9] Design a 1/8 (1/9) frequency divider using 0.18um CMOS technology that works at 1.6Ghz clock frequency.

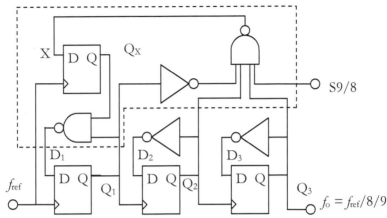

[8.10] Design a 3 bit up/down counter with clear using CMOS gates.

[8.11] Design a 3.2 Ghz divide-by-2 circuit based on the following circuit structure using the 0.18um CMOS process technology

[8.12] Design a 3.2Ghz I/Q clock generation circuit based on the following circuit structure using the 0.18um CMOS process technology

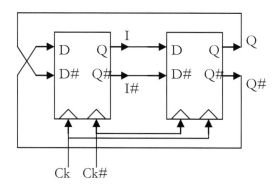

[8.13] Design a VLSI half adder circuit using 0.18um CMOS process technology. Give its truth table.

[8.14] Design a VLSI full adder circuit using 0.18um CMOS process technology. Give its truth table.

8.2 SAMPLE INTERVIEW QUESTIONS

1. Design a decade counter.

2. Explain why VCO can be used as a phase (or time) domain integrator?

3. What is related phase values of 100ps under a 1Ghz, 2Ghz and 10 Ghz system?

4. Explain why frequency divider can serve as a phase scaler.

5. Design a simple circuit based on combinational logic to double the output frequency.

6. What is the function of a D-FF, whose inverted outputs are connected to its input?

7. How do you detect a sequence of "1101" arriving serially from a signal line?

8. Give the truth table for a half adder. Provide a VLSI circuit implementation of a half adder.

9. Draw a Transmission Gate-based D-Latch.

10. Design a Transmission Gate based XOR. Now, how do you convert it to XNOR? (Without inverting the output).

11. How do you detect if two 8-bit signals are same?

12. How to design an 8-bit up/down binary counter?

13. Why PI is a special phase addition/scaling circuit?

9

VLSI PHASE DETECTION CIRCUITS

- VLSI Clock Phase Detectors
- VLSI Data Phase Detectors

Phase detector (PD) circuits are VLSI time-domain signal processing circuit elements to extract the delay phase information from a clock or a data stream. There are two major VLSI phase detector circuit families, including the clock phase detectors (CPD) and the data phase detectors (DPD). VLSI CPD circuits are commonly used to detect the phase information from the clock signals. VLSI DPD circuits, on the other hand, are used to extract the phase information from the random data signal.

VLSI CPD and DPD circuits are key circuit elements of VLSI PLL and DLL and high-speed I/O data recovery circuits.

9.1 HOMEWORK AND PROJECT PROBLEMS

[9.1] Show the timing diagram for the RS latch circuit shown in figure below.

Ck1 —[Q1

Ck2 —[Q2

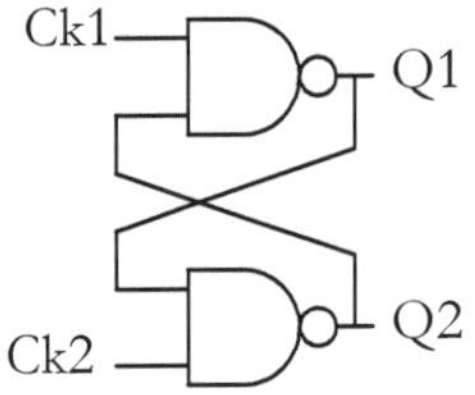

[9.2] Derive the I/O transfer function for the RS latch circuit shown in figure below.

Ck1 —[Q1

Ck2 —[Q2

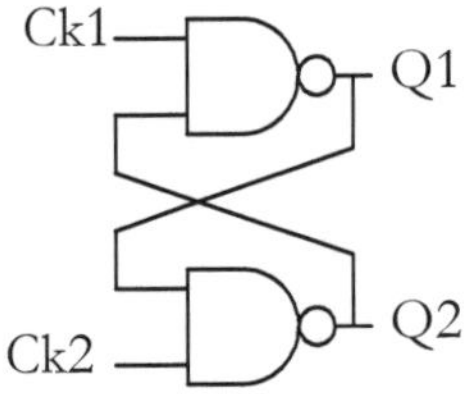

[9.3] Show the state diagram for the RS latch circuit shown in figure below.

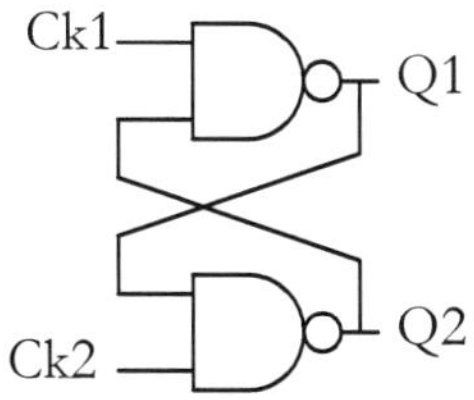

[9.4] Show the timing diagram for the RS latch circuit shown in figure below.

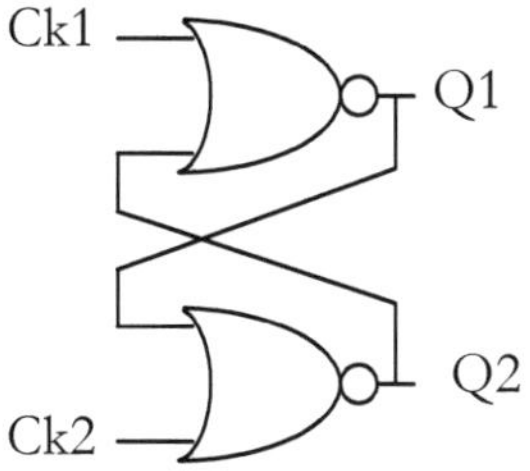

[9.5] Derive the I/O transfer function for the RS latch circuit shown in figure below.

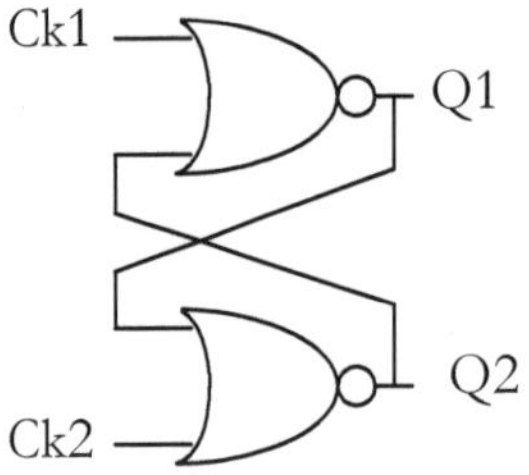

[9.6] Show the state diagram for the RS latch circuit shown in figure below.

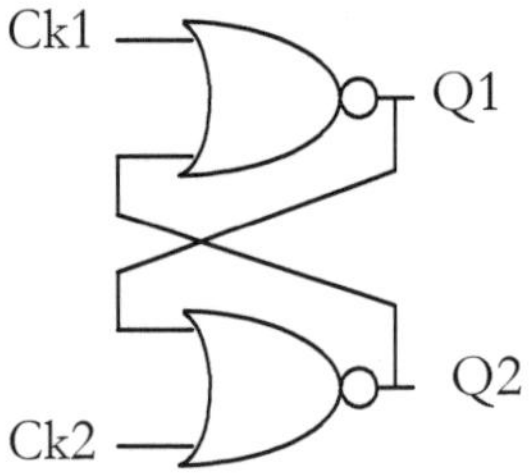

[9.7] Show the timing diagram for the PFD circuit shown in figure below.

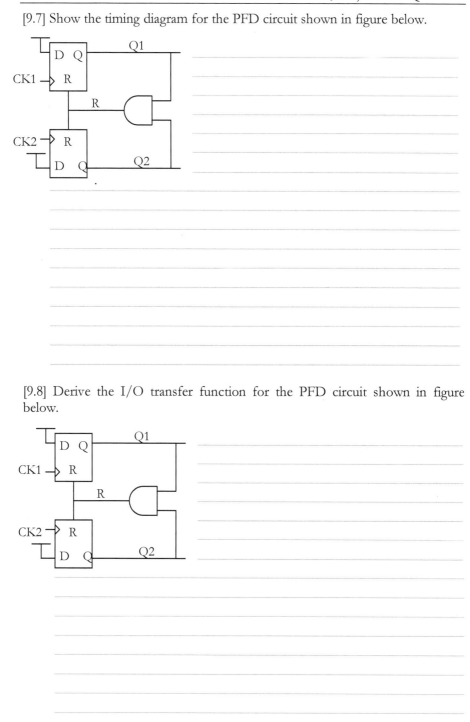

[9.8] Derive the I/O transfer function for the PFD circuit shown in figure below.

[9.9] Show the state diagram for the PFD circuit shown in figure below.

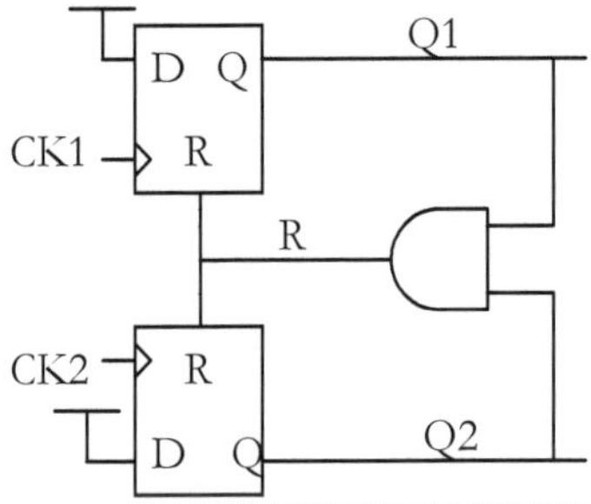

[9.10] Prove the two circuits shown below are equivalent.

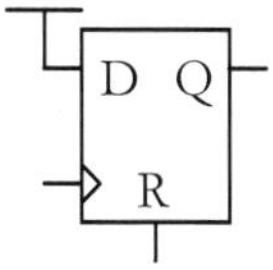

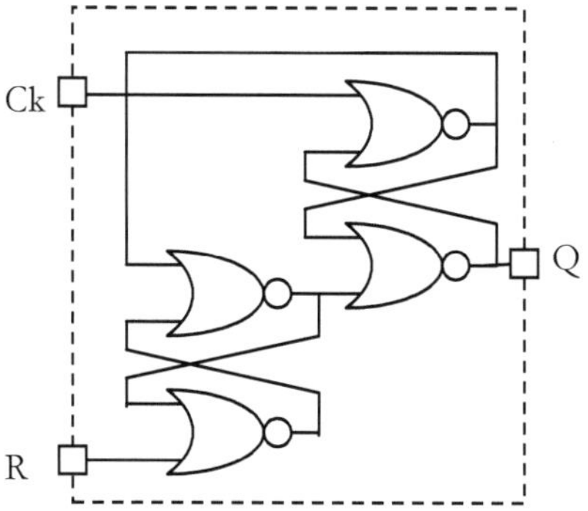

[9.11] Prove the two circuits shown below are equivalent.

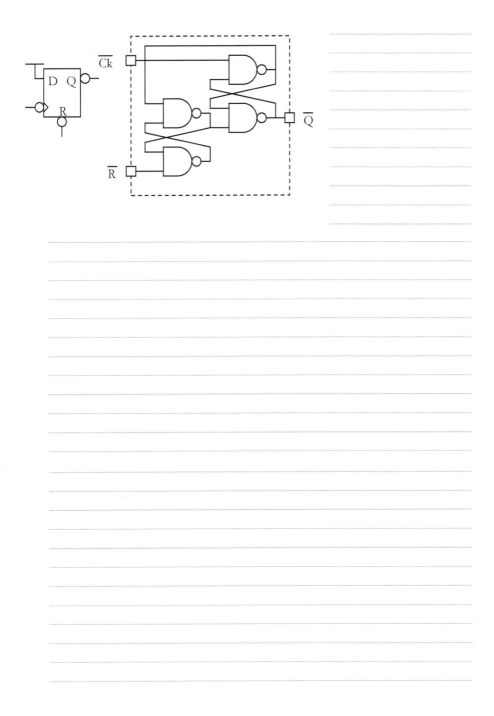

[9.12] Design a PFD using above VLSI circuit to gate level.

[9.13] Determine the input to output transfer function of the VLSI data phase detector circuit shown in figure below.

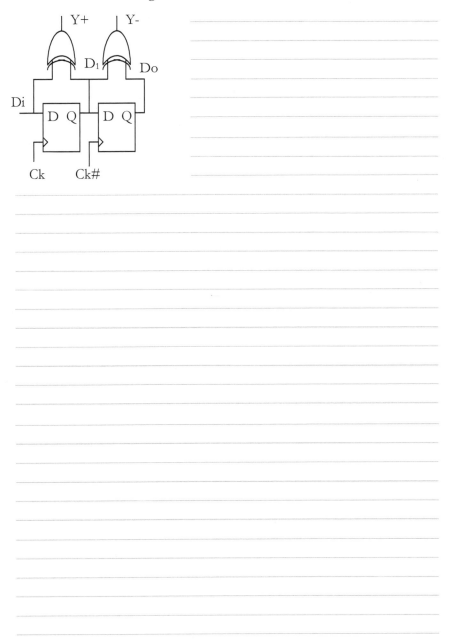

[9.14] Determine the input to output transfer function of the VLSI data phase detector circuit shown in figure below.

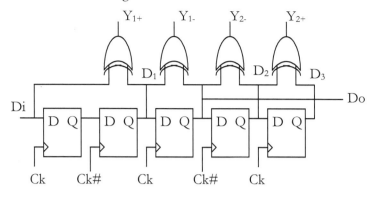

[9.15] Determine the input to output transfer function of the VLSI data phase detector circuit shown in figure below.

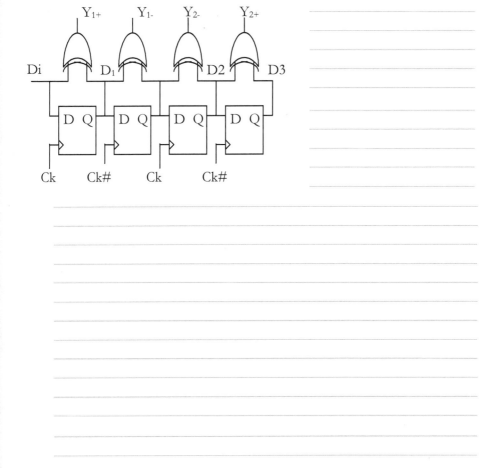

[9.16] Determine the input to output transfer function of the VLSI data phase detector circuit shown in figure below.

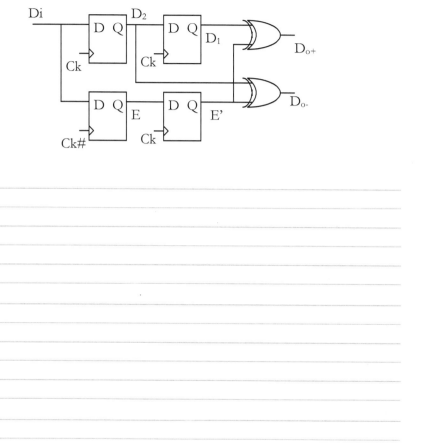

9.2 SAMPLE INTERVIEW QUESTIONS

1. In a PLL, what elements (like XOR gates or Flip-flops) can be used to design the phase detector?

2. Describe the key characteristics of the analog multiplier based phase detector circuit.

3. Can analog multiplier based phase detector be used as data phase detector?

4. What is the phase relationship of the two input clock of the AND phase detector at the locked-in condition?

5. What is the phase relationship of the two input clock of the OR phase detector at the locked-in condition?

6. What is the phase relationship of the two input clock of the XOR phase detector at the locked-in condition?

7. Describe the key characteristics of the R-S latch based phase detector circuit.

8. Describe the key characteristics of the phase frequency detector (PFD) circuit.

9. What is the phase relationship of the two input clock of the PFD at the locked-in condition?

10

VLSI PHASE-LOCKED LOOP CIRCUITS

- Phase Domain PLL SFG Model

- PLL Phase Noise Transfer Function

- VLSI PLL Circuit Implementation

Phase-locked loops (PLL) circuits are widely used in VLSI high-speed I/O circuits. Key applications of VLSI PLL circuits include the frequency (clock) synthesis, the phase (delay) synthesis and the delay variation (jitter) filtering. In frequency synthesis applications, PLL circuits are used to translate the off-chip reference clock (usually at lower frequency) to on-chip clocks with frequency for specific I/O circuit data rates. In the phase synthesis applications, the PLL circuits are usually used associated with the data and clock recovery operations in high-speed I/O circuits, where the receiver reference clocks are regenerated based on the phase of the received data stream such that sampling clock are aligned to the center of the received data eye patterns for optimal data capture. In the jitter filtering applications, PLL circuits are used as clock signal conditioning circuits to eliminate the high frequency jitter contents within the reference clocks.

VLSI PLL circuits suffer from various voltage and timing variation effects that can significantly impact the VLSI high-speed I/O circuit performance. Timing uncertainties of PLL circuits are mainly contributed from the jitter and clock feed-through effect of input reference clocks, the added noises from the core circuits of the PLLs, such as the VCO circuits, the phase detector circuits, and the clock distribution circuits. These jitter effects consist of both the random jitter components from the thermal noise effects of the devices and the deterministic jitter components from the cross-talk noises and power supply noises.

10.1 HOMEWORK AND PROJECT PROBLEMS

[10.1] A PLL has a center frequency of 10^5 rad/s, Ko=10^3 rad/s/v, Kd=1V/rad Assume there is no other gain in the loop. Determine the loop bandwidth in the first-order PLL loop configuration.

[10.2] Determine the single-pole, loop-filter pole location to give the closed-loop poles located on 45 degree radials from the origin of the complex frequency plane.

[10.3] In order to produce poles at 45 degree to the axis, we add a loop filter pole at $\omega 1$ where $\omega 1 = 2 K = 2000$ rad/s, then, the filter transfer function becomes, $F(s) = 1/ (1+ s/\omega 1)$. Why do we choose $\omega 1 = 2 K$?

[10.4] Explain PLL block level functioning and at which point jitter is observed in PLL?

[10.5] For the PLL circuit shown in figure below, derive the input to output phase transfer function.

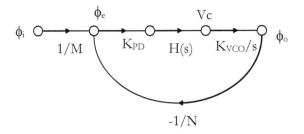

[10.6] For the PLL circuit shown in figure below, derive the input to output phase transfer function.

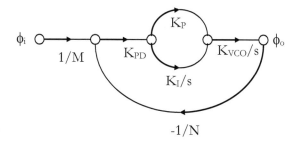

[10.7] Derive the noise transfer functions of the PLL circuit related to the noise voltage input respectively.

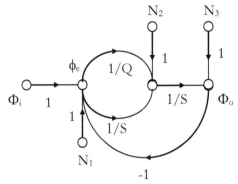

[10.8] Simplify the following PLL model directly in SFG transformation to the prototype form. Calculate the PLL and BW and Q-factor.

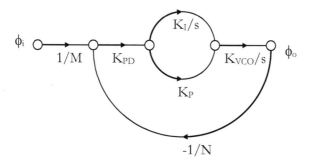

[10.9] Design a 4-phase ring oscillator PLL based on 0.18um CMOS process for 100Mhz input clock reference and 1.6Ghz output frequency.

[10.10] Derive the phase transfer function for the PLL circuit shown in figure below using the given circuit parameters. Derive the PLL bandwidth and quality factor (dumping factor) of this PLL.

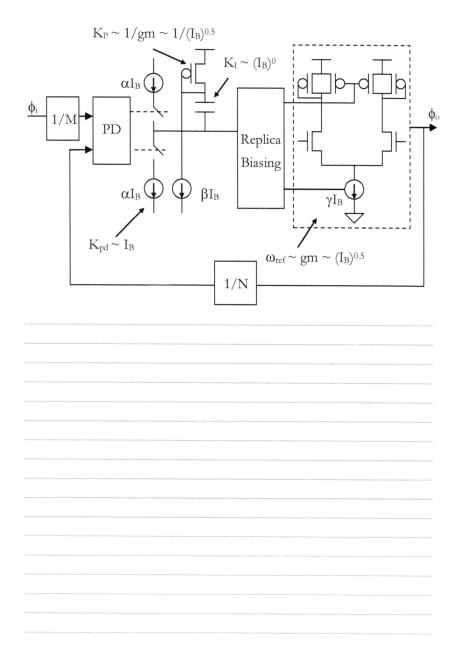

10.2 SAMPLE INTERVIEW QUESTIONS

1. Why the bandwidth of the PLL loop is usually selected to be less than 1/10 of the input clock reference to the phase detector?

2. What is an integer-N PLL? What is a fractional-N PLL?

3. What are the advantages and disadvantages of the fractional-N PLL?

4. What are the key benefits of the VLSI LC-PLL versus the ring oscillator based PLL?

5. What does the SSC mean?

6. Why higher PLL BW is usually desired for ring oscillator PLL versus the LC PLL?

7. Higher or lower PLL bandwidth is desired when a crystal oscillator is used as PLL reference versus a clock chip as clock reference?

8. If the dominant phase noise of the PLL is contributed from the input reference, higher or lower PLL bandwidth is usually desired?

12. If the domain phase noise of a PLL is contributed from the VCO, higher or lower PLL bandwidth is usually desired?

13. What is clock gating? How and why it is done.

14. Why we usually cascade lower division ratio PLLs to achieve very high division ratio PLL?

11

VLSI DELAY-LOCKED LOOP CIRCUITS

- Phase Domain DLL SFG Model
- DLL Phase Noise Transfer Function
- VLSI DLL Circuit Implementation

VLSI delay-locked loop (DLL) circuits are a special family of VLSI phase-locked loop (PLL) circuits that employ the VLSI VCDL circuits. The feedback loop in the DLL circuit ensures the absolute delay time of the VCDL is adaptively locked to a reference delay time independent of the PVT conditions.

VLSI DLL circuits are widely used in VLSI high-speed I/O circuits for clock and data path delay compensation, for clock phase manipulation, for multi-phase clock generation, and for clock multiplication.

There has been a great deal of interests in various DLL circuits in the recent years for data and clock signal delay compensation in VLSI high-speed I/O circuit applications that only requires the phase or delay adjustment and not for frequency synthesis. In these applications, DLL circuits are used as alternative to PLL circuits to provide precise spaced timing edge even in the presence of PVT variations. DLL circuits are preferred circuits in those applications because their unconditional stability, lower phase-error accumulation effects, and fast locking times.

11.1 HOMEWORK AND PROJECT PROBLEMS

[11.1] For the VLSI DLL circuit shown in figure below, N = 10 and M = 15. Assuming the input reference signal is a 1Ghz clock (i.e. reference time period is T = 1ns). What is the delay time from x to y when the circuit is normal operation model?

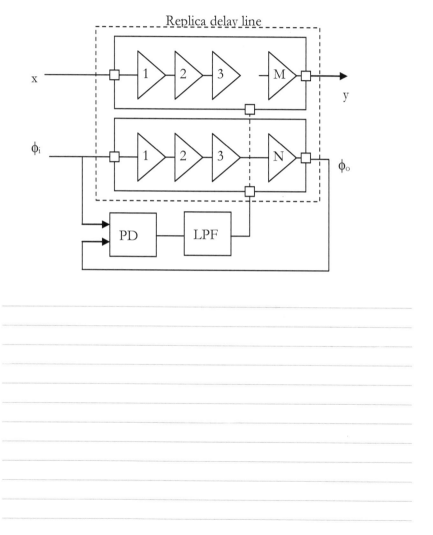

[11.2] Derive the delay time transfer function for the first-order DLL shown in figure below.

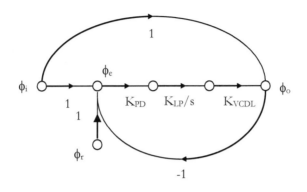

[11.3] Derive the delay time transfer function for the second-order DLL shown in figure below.

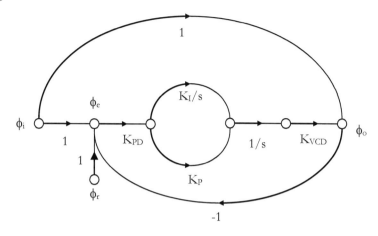

[11.4] Design a 4-phase clock generation DLL based on 0.18um CMOS process for 1.6Ghz clock operation.

[11.5] Explain how this digital DLL works.

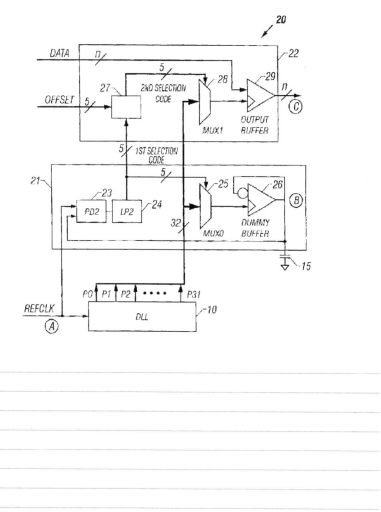

11.2 SAMPLE INTERVIEW QUESTIONS

1. Why PFD might not be suitable for detecting harmonic locking condition in DLL circuit?

2. What is likely the order of the DLL with a first order loop filter?

3. What are the major benefits and limitations of DLL compared with PLL circuit.

12

VLSI ON-CHIP IMPEDANCE TERMINATION CIRCUITS

- Matching of High-Speed I/O Channel
- Programmable On-Chip Termination Circuit
- Adaptive On-Chip Termination Circuit

VLSI high-speed I/O transmitter and receiver circuits are physically located in separate chips within the electronic systems, connected through an electrical channel, usually consisting of the chip package, binding wires, connectors, and motherboard traces. Proper termination of high-speed I/O channel is very critical since the channel reflection and ringing induced jitters are among the major sources of VLSI high-speed I/O circuit delay uncertainty effects. Most VLSI high-speed I/O circuit standards have specific channel characteristic impedances with 50Ω as a typically used impedance value.

Ringing effects may occur in improperly terminated short channels, where the total length of all stubs on a line is not short enough or there is a discontinuity in impedance, such as a connector, in the channel. Ringing effects may cause false signal edge due to signal amplitude variation. Reflections, on the other hand, usually occur in improperly terminated long channels. Reflection effects caused timing problems due to the inherent property of voltage doubling on an open transmission line. Both the ringing and reflection effects can be effectively minimized using the proper impedance termination and slew rate control circuit techniques.

12.1 HOMEWORK AND PROJECT PROBLEMS

[12.1] If a VLSI high-speed I/O channel can be modeled using a RLC network as shown in figure below. Determine the expression for Rtx to eliminate the ringing at the output node.

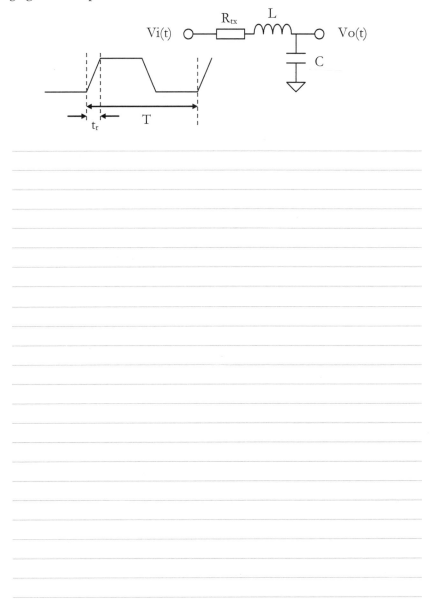

[12.2] If a VLSI high-speed I/O channel can be modeled using a RLC network as shown in figure below. Determine the expression for R_{RX} to eliminate the ringing at the output node.

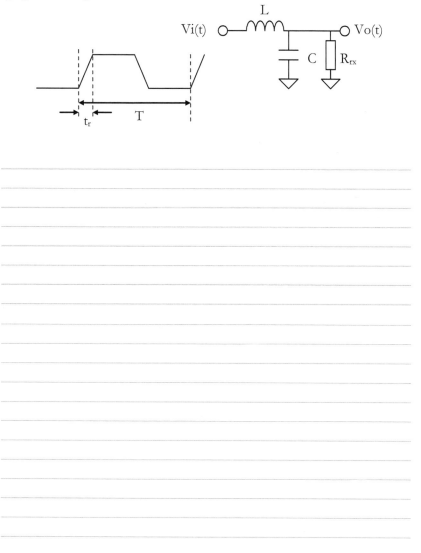

[12.3] If a VLSI high-speed I/O channel can be modeled using a RLC network as shown in figure below. Determine the expression for Rtx and R_{rx} to eliminate the ringing at the output node.

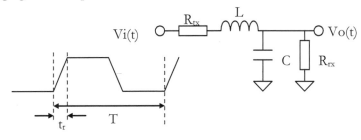

[12.4] Find the equivalent differential and common-mode output impedance of the CML buffer circuit shown in figure below. Please ignore the parasitic capacitance of the circuit in this problem.

[12.5] Find the equivalent differential and common-mode output impedance of the CML buffer circuit shown in figure below, where C is the equivalent capacitance of the two output nodes. Please ignore other parasitic capacitance effects in this problem.

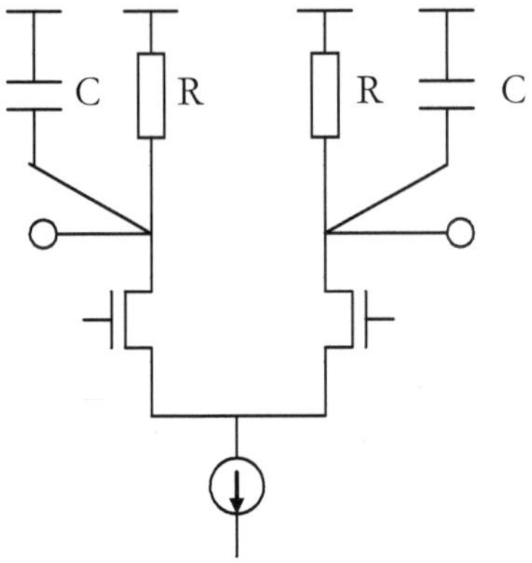

[12.6] Find the equivalent differential and common-mode output impedance of the CML buffer circuit shown in figure below, where C is the equivalent Vcc refereed node capacitance of the node, C1 is the differential node capacitance of the nodes. Please ignore other parasitic capacitance effects in this problem.

[12.7] Find the equivalent differential and common-mode output impedance of the CML buffer circuit shown in figure below, where Co and Ro are the equivalent node capacitance and resistance at tail current node. Please ignore other parasitic capacitance effects in this problem.

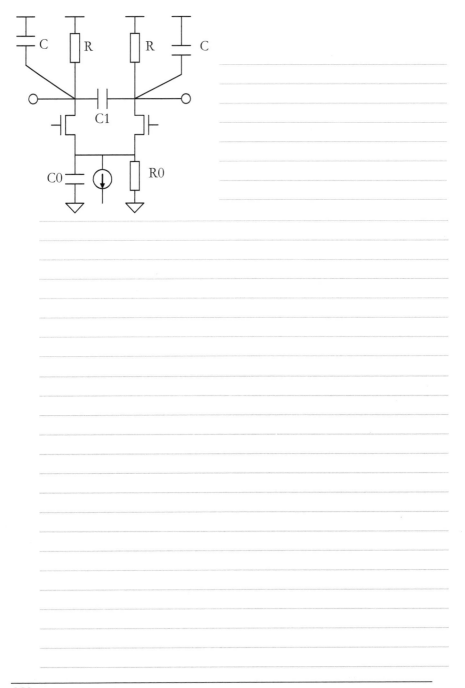

[12.8] Find the equivalent differential and common-mode output impedance of the CML buffer circuit shown in figure below. Please ignore parasitic capacitance in this problem.

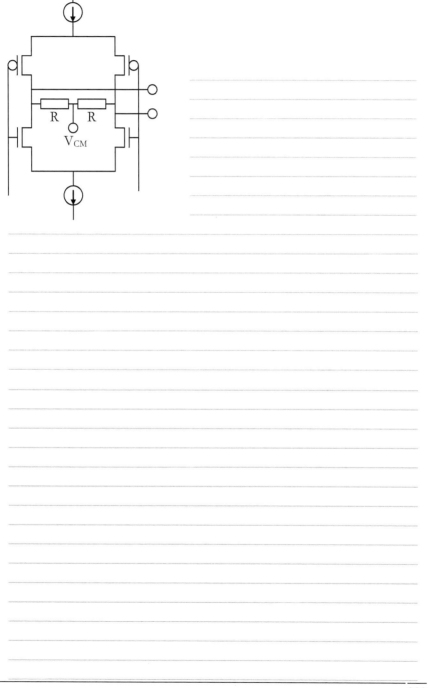

[12.9] Design a 50Ω VLSI adaptive on-chip termination circuit based on the following circuit architecture using the 0.18um CMOS technology. Assuming Vcc = 1.8V.

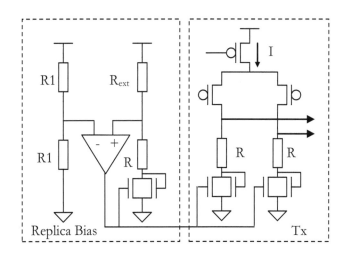

[12.10] Design a 50Ω VLSI adaptive on-chip termination circuit based on the following circuit architecture using the 0.18um CMOS technology. Assuming Vcc = 1.8V.

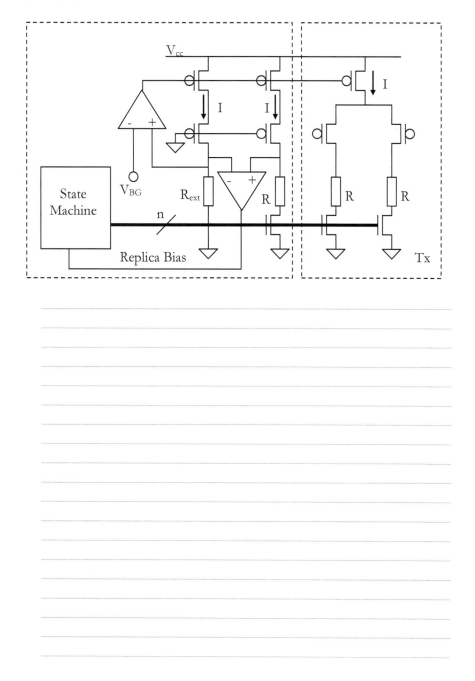

[12.11] Design a 50Ω VLSI adaptive on-chip termination circuit based on the following circuit architecture using the 0.18um CMOS technology. Assuming Vcc = 1.8V.

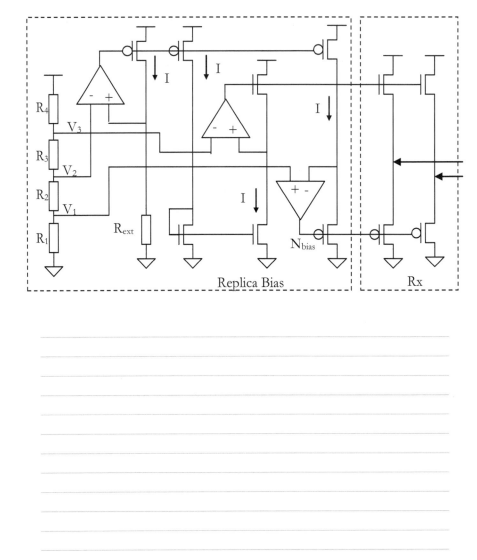

Replica Bias   Rx

[12.12] Assuming that 50Ω+/-20% impedance is required in the high-speed I/O spec (such as PCI-Express) and the VLSI on-chip resistance can change by +/-50% under the PVT corners. What is the minimal control bit in the on-chip resistor tuning circuit if the resistance values are linearly coded?

[12.13] Assuming that 50Ω+/-20% impedance is required in the high-speed I/O spec (such as PCI-Express) and the VLSI on-chip resistance can change by +/-50% under the PVT corners. What is the minimal control bit in the on-chip resistor tuning circuit if the resistance values can be non-linearly coded?

[12.14] Explain how the Tx impedance control circuit shown below works.

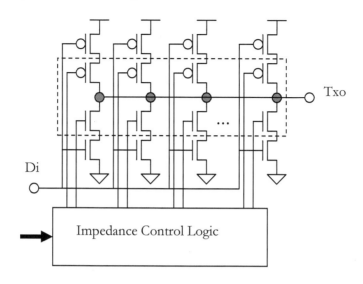

12.2 SAMPLE INTERVIEW QUESTIONS

1. What is the Cpad (pad capacitor) of VLSI high-speed I/O Tx and Rx? Why Cpad is an important SI parameter in high data rate high-speed I/O circuits?

2. What's the advantage and disadvantage of using active device as termination resistor versus using passive devices?

3. Why do we need to terminate high-speed I/O channel?

4. When adaptive on-chip impedance termination is needed?

5. What are the key benefits of on-chip impedance termination versus off-chip impedance termination?

6. Which type of channel (long channel or short channel) is likely more impacted by the termination imperfection?

7. How the S11 at the Rx or Tx is defined in the high-speed I/O circuits?

8. What are the commonly used high-speed I/O impedance termination schemes?

9. What are the commonly used ways to eliminate high-speed I/O ringing effects?

10. What are the benefits and penalties to terminate VLSI high-speed I/O in bother Tx and Rx?

13

VLSI HIGH-SPEED I/O EQUALIZATION CIRCUITS

- Channel Bandwidth Limitation and ISI Effects

- Frequency Domain Model of Channel

- Time-Domain Model of Channel

- Frequency-Domain Approach of Equalization

- Time-Domain Approach of Equalization

Equalization circuits are typically used in VLSI high-speed I/O to minimize the inter-symbol interference (ISI) effects. VLSI equalization circuits can be implemented either in the digital or the analog domains, in the transmitter or in the receiver circuits, using the feed-forward or feedback configurations, or using the linear or non-linear circuits. Practical VLSI high-speed I/O circuits usually combine multiple equalization techniques to achieve better circuit performances.

High-speed I/O equalization circuits can be implemented based on either time-domain or frequency-domain circuit approaches. Time-domain approaches rely on the solution to minimize the spread of the data symbol propagate through the high-speed I/O channel. Frequency-domain approaches, on the other hand, rely on the solution to improve the channel bandwidth.

Transmitter equalization based on the de-emphasis techniques offer a simple VLSI high-speed I/O circuit equalization solution. However transmitter equalizations usually suffer from maximum signal swing limitation and therefore are not suitable for highly loss channels. Both linear and non-linear receiver equalization techniques, such as the feedforward equalizers (FFEs) and decision feedback equalizers (DFEs), are typically used in such design cases.

The adaptive equalization techniques can be used to compensate for unknown or time-varying channel responses. Such equalization techniques are usually based on certain adaptive control algorithms, such as the least-mean-square (LMS), the sign-sign LMS (SS-LMS), etc.

13.1 HOMEWORK AND PROJECT PROBLEMS

[13.1] A normalized (i.e. T = 1) VLSI high-speed I/O channel can be modeled using a RC network as shown in figure.

$$R = 1$$

$V_i(t)$ ⃝———▭———⊤——⃝ $V_o(t)$

$C = 2$

Derive the ISI induced normalized DJ for the circuit that is related to the following repeating data pattern. Verify your result using the circuit simulation.

(a) {01111111111};

(b) {0000000000101111111111};

(c) $\{01\}$;

(d) $\{001\}$;

(e) $\{001011\}$;

(f) {0001};

(g) {00010111};

(h) {00001};

[13.2] A normalized (i.e. T = 1) VLSI high-speed I/O channel can be modeled using a RC network as shown in figure.

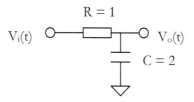

Generate an approximated 5-tap FIR model for this channel. Determine the parameters $\{Y0, Y1, Y2, Y3, Y4\}$ for your model.

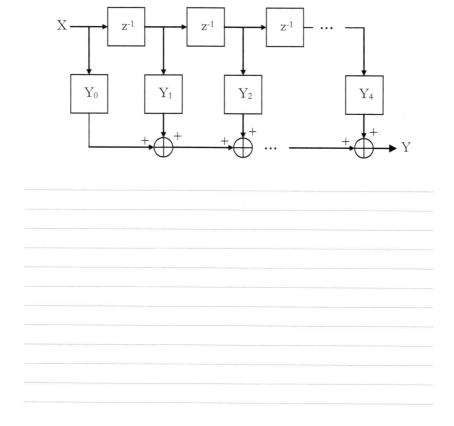

[13.3] Compare the channel frequency responses of two models in above problem.

[13.4] Compare the channel time domain responses (Impulse and Step responses) of two models in problem [13.4].

[13.5] The FIR model of a VLSI equalizer circuit is shown in figure below. Calculate and plot the impulse and step responses of the circuit. Assuming $\alpha = 3/4$ and the clock period is normalized to $T = 1$.

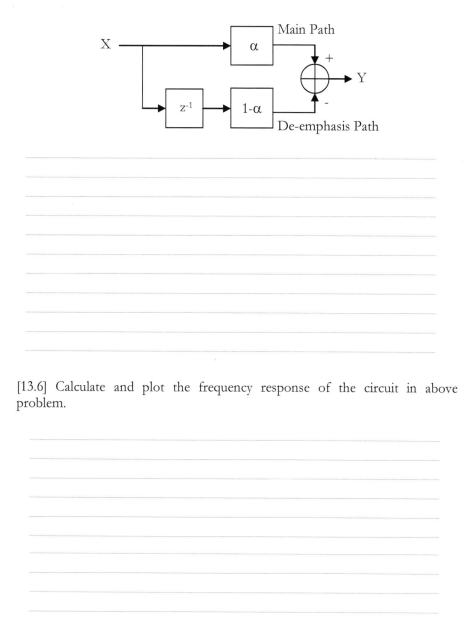

[13.6] Calculate and plot the frequency response of the circuit in above problem.

[13.7] Determine the optimal parameter α for above equalizer that is used to equalize the channel given in [13.2]. Find and plot the impulse responses of un-equalized and equalized channel.

[13.8] For the optimized parameter α in above problem, simulate the eye diagrams of the un-equalized and equalized channel output with PRS data pattern.

[13.9] For the optimized parameter α in above problem, simulate for the frequency responses of the un-equalized and equalized channel output with PRS data pattern.

[13.10] Design a normalized CTLE for the channel given in [13.2] based on the circuit structure shown in below. Determine the normalized R and C value.

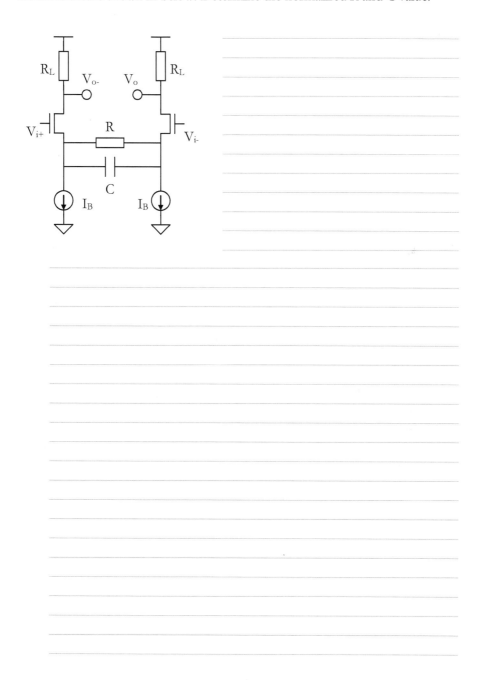

[13.11] Prove your design in [13.10] using simulation based on 0.18um CMOS technology. Please scale the frequency of your design from T = 1 to T = 1ns.

[13.12] Design a normalized CTLE for the channel given in [13.2] based on the circuit structure shown in below. Determine the normalized R, L and C value.

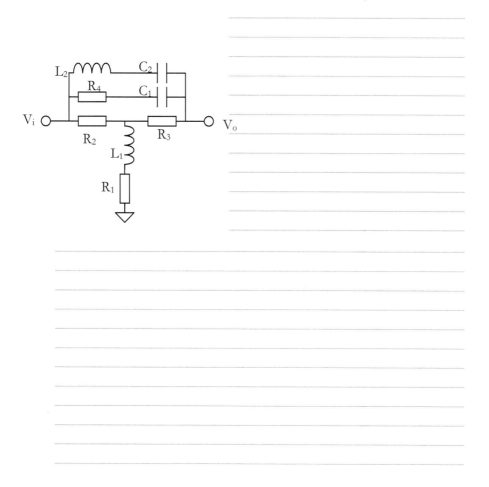

[13.13] Derive the z-domain transfer for the following Tx equalizer circuit.

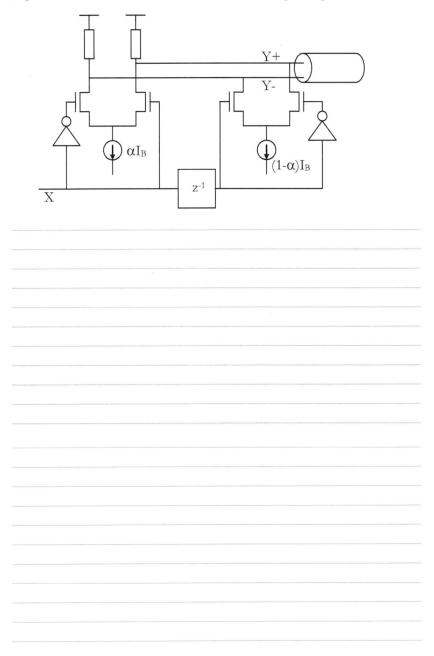

[13.14] Design a 3.5dB VLSI transmitter based on above transmitter architecture for USB2.0 data rate using 0.18um CMOS technology.

[13.15] Derive the transfer function for the following Tx equalizer.

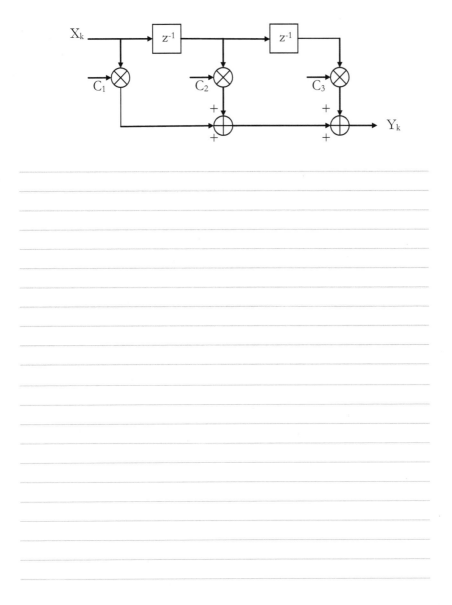

[13.16] Derive the transfer function for the following equalizer.

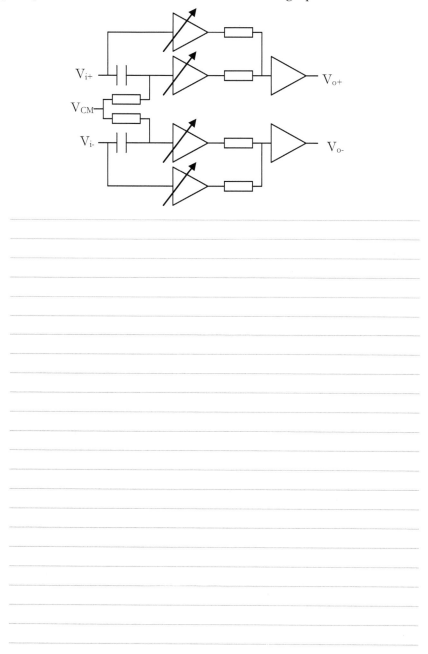

[13.17] Derive the transfer function for the following equalizer.

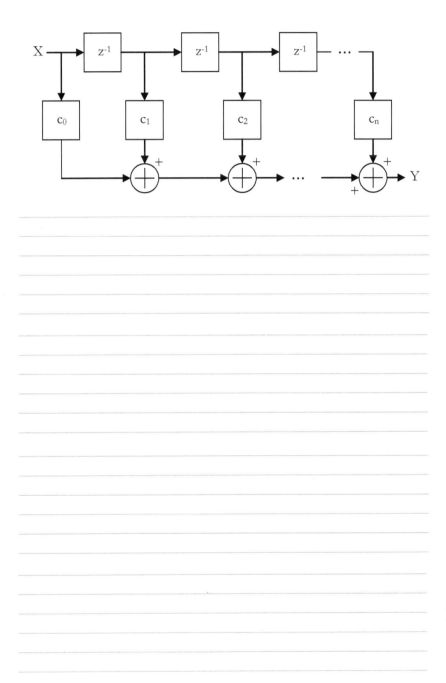

13.2 SAMPLE INTERVIEW QUESTIONS

1. What are the major contributors of frequency dependent channel response?

2. What are meant by pre-emphasis and de-emphasis?

3. What are the major differences of the frequency- and time-domain approaches of the VLSI high-speed I/O channel modeling?

4. What are the major differences between the CTLE (continuous-time equalizer) and DTLE (discreet-time linear equalizer)?

5. Why Rx equalizer is critical for VLIS high-speed I/O circuit with highly loss channel?

6. List commonly used Rx equalizer approaches you know.

14

BASIC VLSI HIGH-SPEED I/O CIRCUIT ARCHITECTURES

- Common Clock I/O Circuits

- Forward Clock I/O Circuits

- Embedded Clock I/O Circuits

There are three basic I/O circuit architectures, including the common clock I/O (also known as synchronous I/O), the forwarded clock I/O (also known as source synchronous I/O) and the embedded clock I/O (also known as data recovery I/O) circuits.

The common clock I/O circuits are based on the concept of electrically matched clock distribution to both the transmitter and receiver circuits from the same reference clock source. This I/O circuit type offers the timing plan simplicity. However it suffers from the limitation of lower achievable delay uncertainty control.

The forwarded clock I/O circuits are based on the concept of physical channel replica for sending the reference clock together in parallel with data from transmitter to the receiver circuit. This I/O architecture can be effectively used to minimize the delay variation in the data transmission. The drawback of the I/O architecture is the cost associated with the replica channel.

The embedded clock I/O circuits are based on the concept of clock recovery circuits for adaptive delay compensation of the transmitter, channel and reference clocks. Such I/O architecture offers effective delay uncertainty control at lower cost. The penalty of such I/O circuit is the increased circuit design complexity.

14.1 HOMEWORK AND PROJECT PROBLEMS

[14.1] A VLSI common clock I/O circuit is shown in figure below. Derive the timing constraint equation for the timing parameters for such circuit.

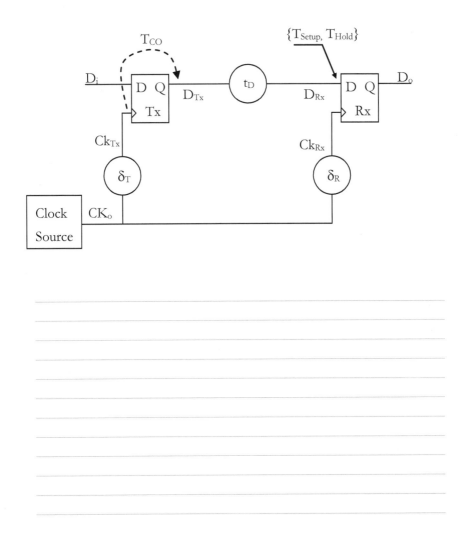

[14.2] For the VLSI common clock I/O circuit shown in figure below, derive the expression of the track phase error E.

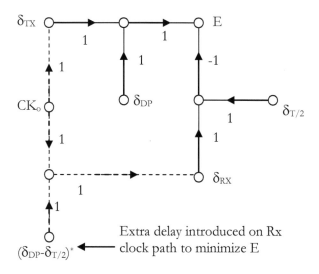

δ_{TX}

E

1 1

1

1 -1

CK_o

δ_{DP}

$\delta_{T/2}$

1

1 1

1

δ_{RX}

1

1

$(\delta_{DP}-\delta_{T/2})^*$ ← Extra delay introduced on Rx clock path to minimize E

[14.3] For the VLSI common clock I/O circuit shown in figure below, if the only delay variation is caused by the jitter amplification of the channel that accounts for +/-20ps delay variation. Determine the maximum allowable data rate of such high-speed I/O circuit.

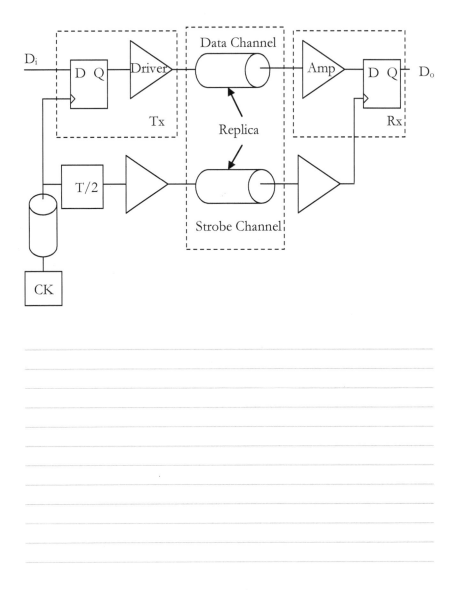

[14.4] For the VLSI common clock I/O circuit shown in figure below, derive the expression of the track phase error E.

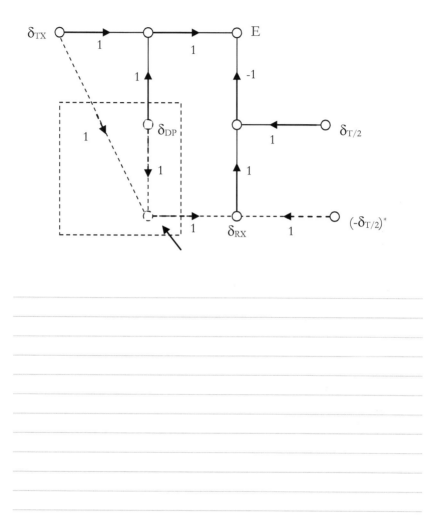

[14.5] For the VLSI common clock I/O circuit shown in figure below, if the delay variation caused by the jitter amplification of the channel is reduced to +/-5ps. In addition, there is about 5% DCD in the clock channel. Determine the maximum allowable data rate of such high-speed I/O circuit.

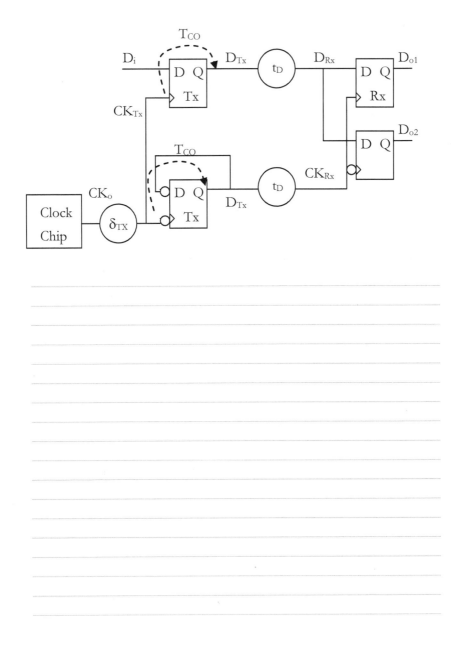

[14.6] For the VLSI common clock I/O circuit shown in figure below, derive the expression of the track phase error E.

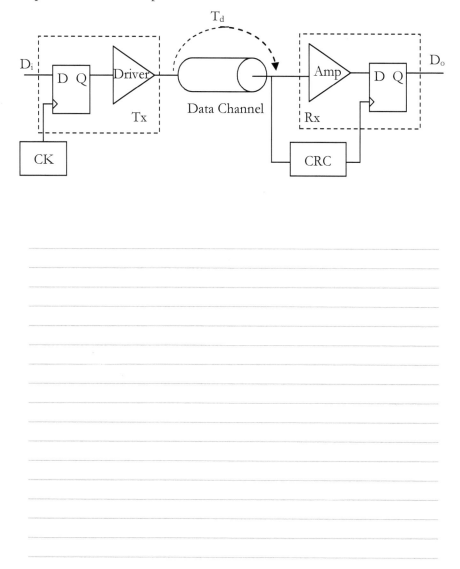

[14.7] For the VLSI common clock I/O circuit shown in figure below, derive the expression of the track phase error E.

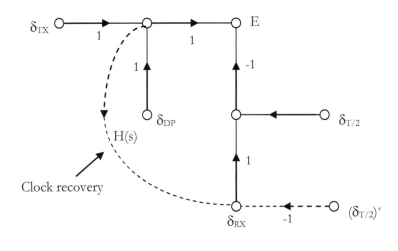

14.2 SAMPLE INTERVIEW QUESTIONS

1. What are the three basic VLSI I/O circuit architectures?

2. What is the common clock I/O circuit?

3. What are the key benefits and limitations of the common clock I/O circuits?

4. What is another name of the common clock I/O circuit?

5. What is the forward I/O circuit?

6. What are the key benefits and limitations of the forward clock I/O circuits?

7. What is another name of the forward clock I/O circuit?

8. What is the embedded I/O circuit?

9. What are the key benefits and limitations of the embedded clock I/O circuits?

10. What is another name of the embedded clock I/O circuit?

11. How to reduce the cost of the forward clock I/O circuits?

12. How to minimize the jitter amplification induced error in the forward clock I/O circuits?

13. What is likely the I/O architecture type for the PCI I/O circuits?

14. What is likely the I/O architecture type for the PCI-Express I/O circuits?

15. What is likely the I/O architecture type for the USB2.0 I/O circuits?

16. What is likely the I/O architecture type for the SATA I/O circuits?

17. What is likely the I/O architecture type for the MiPi I/O circuits?

18. What is likely the I/O architecture type for the HDMI I/O circuits?

19. What is likely the I/O architecture type for the DDR I/O circuits?

15

VLSI HIGH-SPEED I/O TRANSMITTER CIRCUITS

- VLSI High-Speed I/O Transmitter Circuit Operation

- Transmitter Driver

- Transmitter Equalizer

- Transmitter Termination

VLSI high-speed I/O transmitter (Tx) circuits are used for signal conditioning for matching the characteristics of high-speed I/O channel such that it can be transmitted effectively. VLSI high-speed I/O transmitters need to meet dedicated high-speed I/O link specifications, such as the signal swing, loading driving capabilities, bandwidths, slew rates, characteristic impedances, and data rate. Practical VLSI high-speed I/O transmitter circuit may typically include the circuit functions such as the parallel-in-serial-out (PISO) circuit, the (adaptive or fixed) impedance termination circuit, the channel equalization, and the pre-driver circuit. VLSI high-speed I/O transmitter circuit may also contain circuit features such as the design for testability (DFT) circuit and the power management (PM) circuit.

VLSI high-speed I/O transmitter circuits can be modeled in time-domain using a synchronization circuit element with the key timing parameter specified by the equivalent transmitter clock to output (TCO) delay.

Practical VLSI high-speed I/O transmitter circuits typically suffer from various delay time variation effects such as the transmitter reference clock delay variations, the data path delay variations. Key design tasks of VLSI high-speed I/O transmitter circuits are to minimize the skew, jitter, and duty-cycle errors of transmitter reference clock buffer circuits, to minimize the transmitter TCO delay time variation under the PVT variations and loading conditions (capacitive loads, transmission-line and PCB trace discontinuities), to minimize the channel reflection, ringing and cross-talk effects through proper control of transmitter output impedance and slew rate, and to minimize the channel ISI effects through effective implementation of the transmitter equalization.

15.1 HOMEWORK AND PROJECT PROBLEMS

[15.1] Identify timing parameters in the following SFG model that is likely determined by the high-speed I/O transmitter circuit.

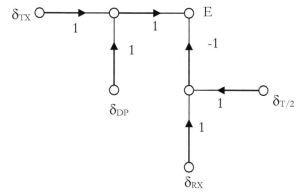

[15.2] Shown below is a VLSI transmitter slew rate control circuit. Please explain how it works.

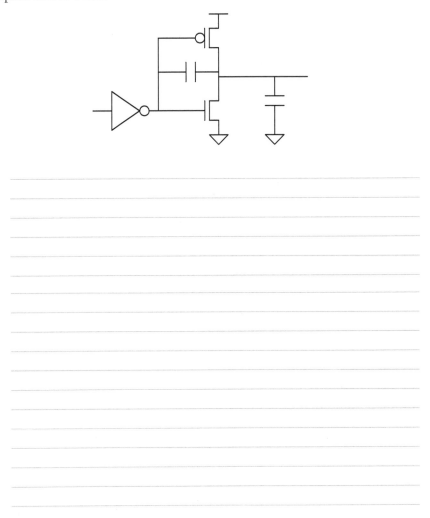

[15.3] For the following circuit structure, what is the steady state signal waveform at X and Y nodes? Assuming RC >>T

[15.4] For the VLSI high-speed I/O circuit shown below, find the steady state waveform at node X+, X-, Y+ and Y-.

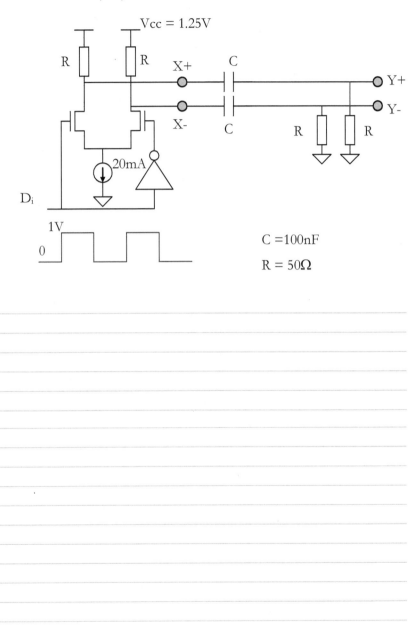

[15.5] Design a high-speed I/O transmitter circuit based on the following circuit structure for the first generation PCI-express link using 0.18um CMOS process technology.

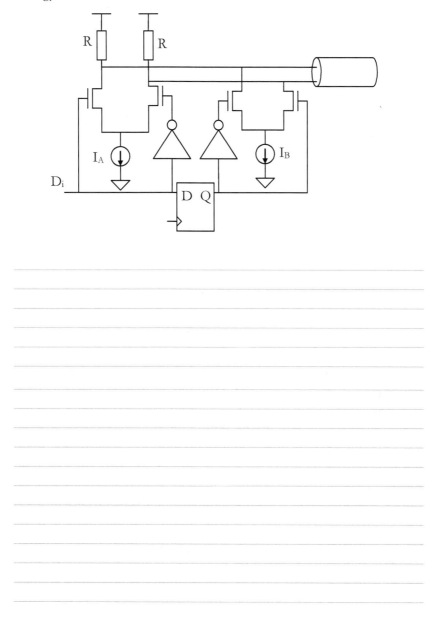

[15.6] Shown below is a VLSI voltage mode Tx circuit architecture. Explain the key benefits and drawbacks of such transmitter circuit versus the current mode Tx circuit architecture as shown in figure above.

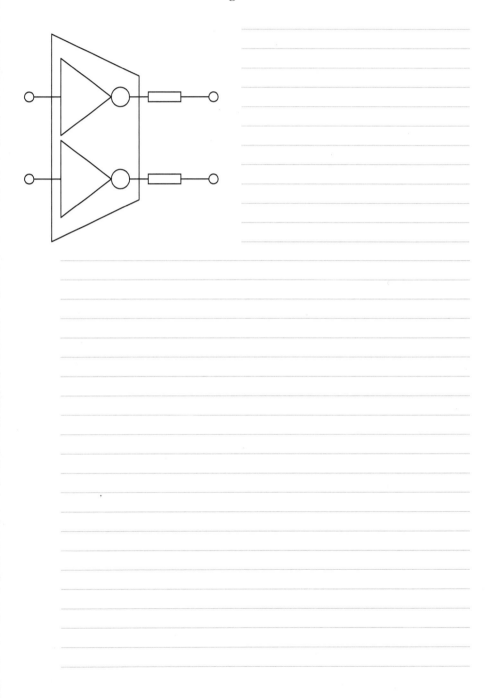

[15.7] Which of the following circuit blocks is likely NOT part of VLSI high-speed I/O Tx circuits?

a) Termination circuit

b) Equalizer circuit

c) SIPO circuit

d) Clock driver circuit

[15.8] Generate a timing diagram for the PISO circuit shown in figure below.

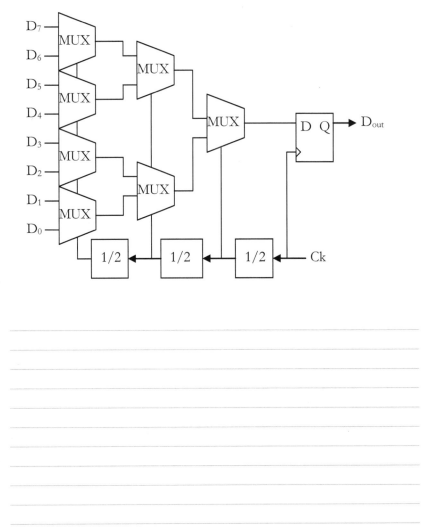

[15.9] Design a VLSI PISO circuit based on above circuit architecture for 3.2Gbps data rate using the 0.18um CMOS process technology.

[15.10] Generate a timing diagram for the PISO circuit shown in figure below.

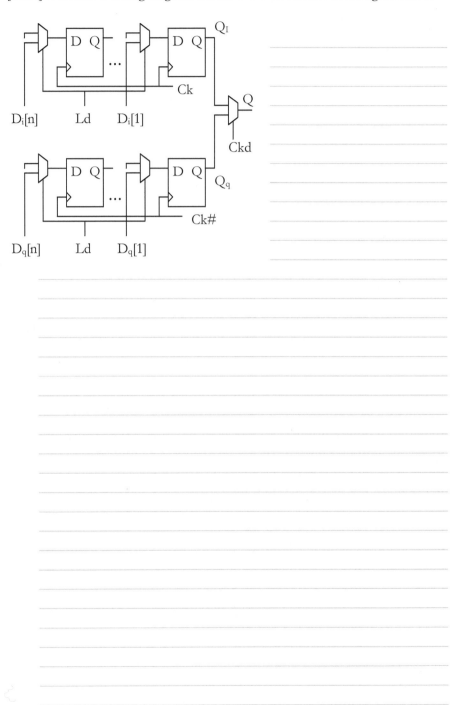

[15.11] Design a VLSI PISO circuit based on above circuit architecture for 1.6Gbps data rate using the 0.18um CMOS process technology.

[15.12] What functions does the VLSI high-speed I/O transmitter given below support? Explain how it works.

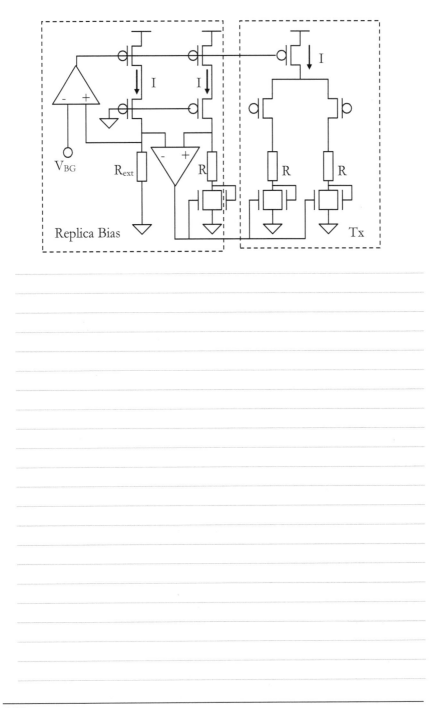

[15.13] Design a VLSI transmitter for USB2.0 based on above circuit architecture using the 0.18um CMOS technology.

[15.14] What functions does the VLSI high-speed I/O transmitter given below support? Explain how it works.

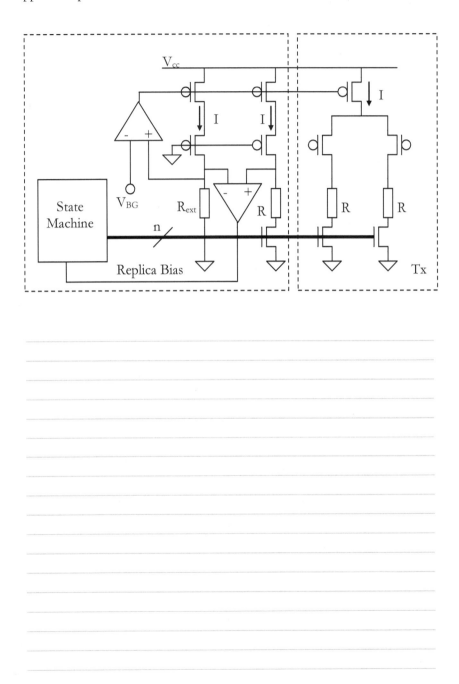

[15.15] Design a VLSI transmitter for USB2.0 based on above circuit architecture using the 0.18um CMOS technology.

15.2 SAMPLE INTERVIEW QUESTIONS

1. What are the key functions of a VLSI high-speed I/O transmitter circuit?

2. Why a digital FF cannot be directly used as high-speed I/O transmitter?

3. Why a CMOS inverter cannot be directly used as high-speed I/O Tx driver?

4. What does LVDS mean?

5. Describe the function of high-speed I/O Tx termination circuit.

6. Why adaptive termination circuit is commonly used in VLSI high-speed I/O transmitter?

7. What is likely the Tx output impedance of a PCI-Express I/O?

8. What is likely the Tx output impedance of a SATA I/O?

9. What is likely the Tx output impedance of a USB2.0 I/O?

10. What is likely the Tx output impedance of a MIPI I/O?

11. What is likely the Tx output impedance of a HDMI I/O?

12. Describe the function of high-speed I/O Tx PISO circuit.

13. Describe the function of high-speed I/O Tx equalizer circuit.

14. What are the major differences between the voltage-mode and current-mode Tx circuits?

15. Describe the function of high-speed I/O Tx slew rate control circuit.

16. Why slew rate control is sometimes needed in VLSI high-speed I/O circuits?

17. Why slew rate control circuit alone is less effective for high data rate I/O circuits?

18. In the design of a large driver, why do we prefer to connect small transistors in parallel instead of laying out one transistor with large width?

19. Describe the key Tx parameters that may impact the ringing and reflection of the high-speed I/O circuits?

16

VLSI HIGH-SPEED I/O RECEIVER CIRCUITS

- VLSI High-Speed I/O Receiver AFE Circuits Models

- VLSI Receiver Preamplifier Circuits

- VLSI High-Speed Sampler Circuits

- VLSI Receiver Channel Termination Circuits

- SIPO Circuits

Receiver (Rx) circuit is used in VLSI high-speed I/O circuit to recover the highly distorted input signal stream to ensure quality of data communication. A VLSI high-speed I/O Rx circuit realizes two major circuit operations including the signal pre-conditioning for SNR enhancement and the timing recover for optimal data sampling.

SNR enhancement operations are implemented using circuits, such the Rx termination and Cpad minimization to minimize the reflection and cross-talk effects, the Rx equalization to minimize the channel ISI effects, the wideband pre-amplification and adaptive gain control to enhance signal power, the offset compensation/calibration to minimize the DC noise.

Timing recovery operations are implemented using a data recover circuit (DRC) or clock and data recovery circuit (CDR) loop that adaptively detects the phase of the incoming data phase to generate a Rx sampling clock using a PI, DLL, or PLL to achieve optimal data sampling. For the forward clock I/O circuits, timing recovery is implemented based on replica channel that sends clock reference with data stream, such that delay variations in the data stream can be effective predicted and compensated.

16.1 HOMEWORK AND PROJECT PROBLEMS

[16.1] Identify timing parameters in the following SFG model that might be impacted by the high-speed I/O Rx circuit.

[16.2] Design a VLSI high-speed I/O sense amplifier sample/hold circuit based on the following circuit architecture for 1.6Ghz application based on 0.18um CMOS technology with worst case receiver sensitivity of 25mV (differential pk-pk). Simulate your design for the setup and hold time and TCO for 25mV diff pk-pk signal input.

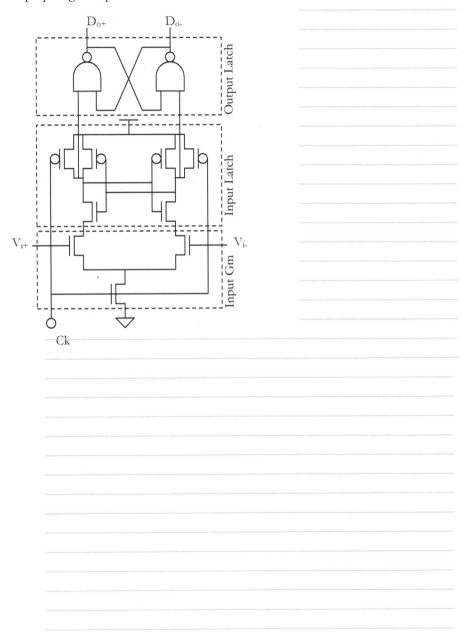

[16.3] Describe how the following SIPO works. Design a 1-to-20 bit SIPO circuit for 1.6Gbps speed using the 0.18um CMOS technology.

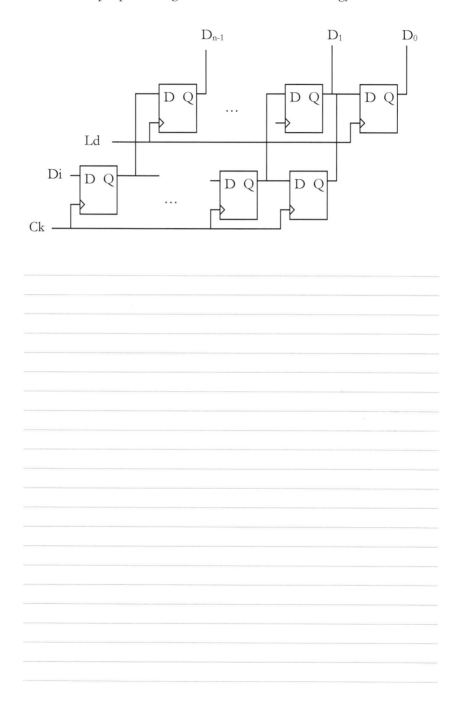

[16.4] Generate a timing diagram for the SIPO circuit shown in figure below.

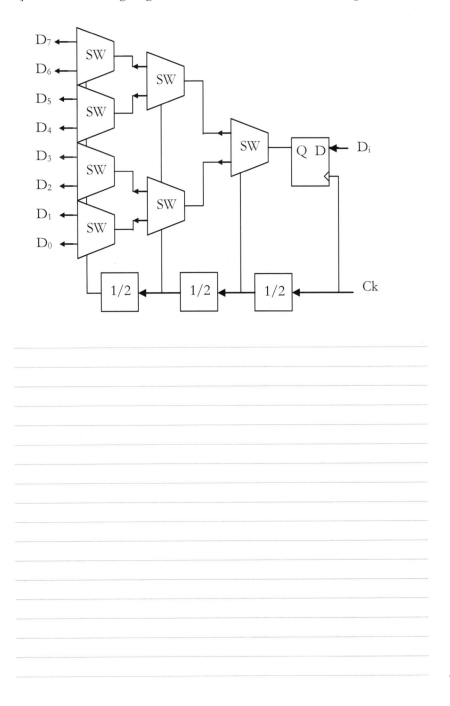

[16.5] Design a VLSI SIPO circuit based on above circuit architecture for 1.6Gbps data rate using the 0.18um CMOS process technology.

[16.4] Which of the following circuit blocks is likely NOT part of VLSI high-speed I/O Rx circuits?

a) Termination circuit

b) Equalizer circuit

c) PISO circuit

d) Clock driver circuit

16.2 SAMPLE INTERVIEW QUESTIONS

1. What are the key functions of a VLSI high-speed I/O Rx circuit?

2. Why a digital FF cannot be directly used as high-speed I/O receiver?

3. What happens if we use an inverter instead of the differential sense amplifier in I/O Rx?

4. Why a CMOS inverter cannot be directly used as high-speed I/O Rx amplifier?

5. Why reflection and x-talk noise have more impact on high-speed I/O Rx than the Tx?

6. Which channel (long channel or short channel) is more sensitive to the imperfection in channel impedance termination?

7. Which channel (long channel or short channel) is more sensitive to the imperfection in channel equalization?

8. Describe the function of high-speed I/O Rx termination circuit.

9. Why adaptive termination circuit is commonly used in VLSI high-speed I/O Rx?

10. What is likely the Rx input impedance of a PCI-Express I/O?

11. What is likely the Rx input impedance of a SATA I/O?

12. What is likely the Rx input impedance of a USB2.0 I/O?

13. What is likely the Rx input impedance of a MIPI I/O?

14. What is likely the Rx input impedance of a HDMI I/O?

15. What is the key function of differential pre-amplifier in high-speed I/O Rx circuit?

16. Explain the working of pre-amplifier in high-speed I/O Rx circuit.

17. Explain the function of high-speed I/O Rx termination circuit.

18. Explain the function of high-speed I/O equalizer circuit.

19. What commonly used high-speed I/O Rx equalizer families?

20. Why offset calibration circuits are commonly used in high-speed I/O Rx circuits and not in the Tx circuits?

21. What happens if we use an inverter instead of the differential amplifier in VLSI high-speed I/O Rx?

22. What is ESD? How to protect Rx from ESD?

23. Explain what is matching in Rx design. What happens if circuit matching does not meet?

24. Explain the need for circuit matching in Rx design.

25. Expand the types of circuit matching in Rx circuit design.

26. How can an inverter work as an amplifier?

27. How can you find the noise margin of a RX circuit?

28. What does it mean by Rx Jitter tolerance?

17

VLSI HIGH-SPEED I/O DATA RECOVERY CIRCUITS

- High-Speed I/O Data Recovery Circuit Operations
- Time-domain DRC SFG Models
- VLSI DRC Implementations
- Receiver Noise Margin Models

Timing recovery operation is among the most important operations of VLSI high-speed I/O circuits. Timing recovery in VLSI embedded clock high-speed I/O circuits is usually realized using the data recovery circuits (DRC) that serves as a special phase-locked circuit to adaptively control the receiver reference clock phase based on the phase information of the received data stream such that the receiver sampling clock aligns to the center of the incoming data eye pattern. VLSI data recovery circuits estimate the most probable timing at which the incoming data stream should be sampled by comparing the estimated timing and the timing information embedded in the data transitions and determine the digital values based on the optimal sampling of signal.

The performance of the data recovery circuit is typically specified by the jitter tolerance and jitter transfer functions. Jitter tolerance is defined as the amplitude of sinusoidal jitter versus frequency a receiver DRC can handle for a given bit error rate (BER). Jitter tolerance represents the ability of the DRC to recover an incoming serial data correctly despite the applied jitter. Jitter tolerance of the DRC is usually specified using a jitter tolerance mask. Jitter transfer function is defined as the ratio of the DRC output jitter of a recovered clock to the input jitter versus frequency. Jitter transfer function represents the gain by which the DRC attenuates or amplifies jitter. Jitter tolerance and transfer function are usually directly related to the DRC loop frequency response. A reasonable loop bandwidth is usually desired for better jitter transfer function characteristics and loop stability.

17.1 HOMEWORK AND PROJECT PROBLEMS

[17.1] Derive the jitter transfer function of the VLSI high-speed I/O data recovery circuit shown in SFG below.

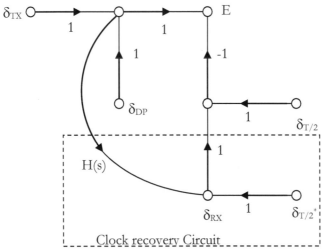

[17.2] Derive the jitter transfer function of the VLSI high-speed I/O data recovery circuit shown in SFG below.

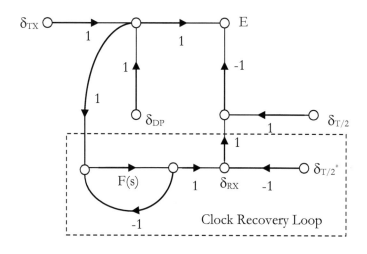

[17.3] Derive the jitter transfer function of the VLSI high-speed I/O data recovery circuit shown in SFG below.

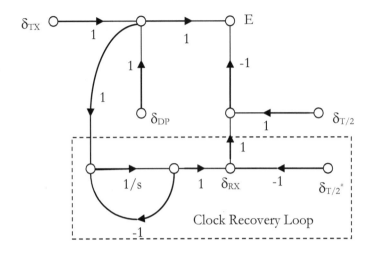

[17.4] Derive the jitter transfer function of the VLSI high-speed I/O data recovery circuit shown in SFG below.

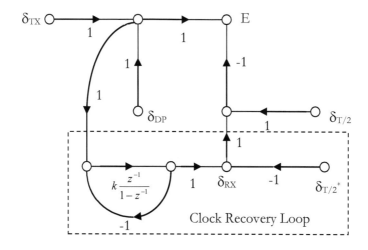

[17.5] Derive the jitter transfer function of the VLSI high-speed I/O data recovery circuit shown in SFG below.

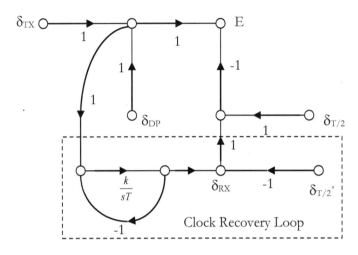

[17.6] Derive the jitter transfer function of the VLSI high-speed I/O data recovery circuit shown in SFG below.

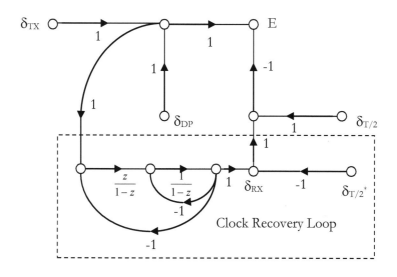

[17.7] Derive the jitter transfer function of the VLSI high-speed I/O data recovery circuit shown in SFG below.

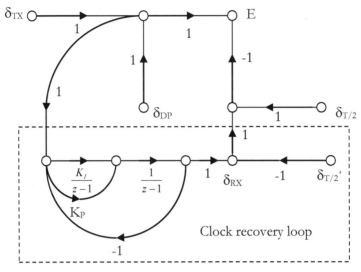

[17.8] For a VLSI high-speed I/O DRC circuit employing a 64 step PI circuit as shown below that updates every four UIs (i.e. update based on averaging of the DPD output the four data bits). Assuming the data is 8B/10B encoded such that in the worst case condition, there is one transition in every five data bits. Derive the maximum allowed transmitter and receiver frequency difference (in PPM).

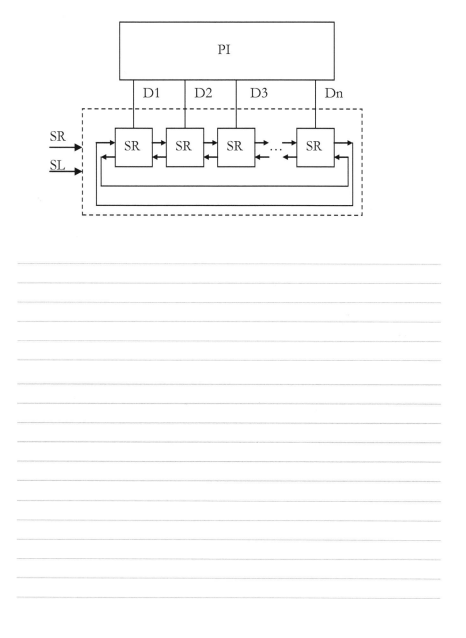

[17.9] For a 5Gbps DRC circuit given in [17.8], what is like the worst case DRC loop bandwidth ω_o?

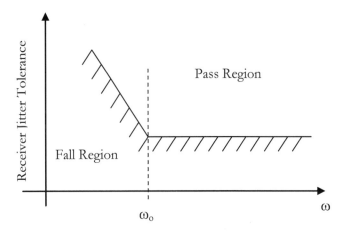

17.2 SAMPLE INTERVIEW QUESTIONS

29. What is the function of data recovery circuit (DRC) in high-speed I/O circuits?

30. Why CDR is another name of DRC?

31. Why DRC is usually not needed in the basic common clock I/O?

32. Why DRC is usually not needed in the basic forward clock I/O?

33. How to determine the bandwidth of the high-speed I/O DRC loop?

34. Why slightly higher DRC bandwidth is desired for better Rx jitter tolerance?

35. What determines the highest allowable DRC loop bandwidth in a high-speed I/O circuits?

36. Why very high DRC loop bandwidth may make the DRC loop not stable?

37. Why a sync circuit block is usually between the DRC and the core circuit?

38. What are the major differences using the PI, the VCDL and VCO in the DRC circuits?

39. Why don't we care the very low frequency jitter component with a VLSI high-speed I/O circuit?

40. Why the raw jitter data of the PLL cannot be used in the VLSI high-speed I/O jitter budgeting? How should the jitter be processed to given a meaningful result in jitter spec?

41. In the VLSI high-speed I/O circuit with 10Mhz DRC loop bandwidth, a 33kHz SSC is used in the clock source to minimize the EMI effect. If the SSC induced peak jitter is 1ns, how large is the impact (in ps) of this SSC induced jitter to this I/O link?

18

VLSI JITTER FILTER CIRCUITS

- VLSI Jitter Transfer Function Model
- Jitter Transfer Function of Filter-Like Circuits
- Duty-Cycle Distortion Models And Correction
- Polyphase Clock Phase Spacing Error And Correction

Timing jitter and phase errors may interact with VLSI circuits. Based on the characteristics of the circuits, jitter may experience frequency selection, amplification, and attenuation effects. Practical VLSI signal processing circuits belong two major circuit families with respect to their interaction with the timing jitters, including the delay (or phase) domain signal processing circuit family and the filter-like signal processing circuit family.

VLSI phase-locked loop (PLL, DLL, DRC, etc) and other time-domain signal processing circuits belong to the first circuit family that are usually directly specified and designed in delay or phase signal forms and their signal transfer functions are the same with their jitter transfer function. VLSI analog filter circuits, clock buffer circuits, interconnects, transmitter circuit, receiver AFE circuits, and the I/O channel belong to the second circuit family that are usually specified in voltage (or current) domain and their signal transfer functions are significantly different from their jitter transfer functions.

Timing jitter of high-speed I/O circuits can be explicitly or implicitly impacted by the phase domain or filter-like VLSI signal processing circuits respectively. The effects can be expressed using their jitter transfer functions.

18.1 HOMEWORK AND PROJECT PROBLEMS

[18.1] A clock signal with introduced jitter can be expressed using the phase modulated sinusoidal equation as:

$$f(t) = A\cos(\omega t + \phi_m \cos(\omega_m t + \alpha) + \delta)$$

Assuming $\phi m \ll 1$ and $\omega_m \ll \omega$.

a) What is the nominal clock frequency of this clock signal in terms of parameters given in the equation?

b) What is the peak-peak, mean, and rms cycle-cycle jitter of this clock in terms of parameters given in the equation?

c) What is the peak-peak, mean, and rms TIE of this clock in terms of parameters given in the equation?

d) What is the peak-peak, mean, and rms period jitter of this clock in terms of parameters given in the equation?

[18.2] Derive the input to output jitter transfer function of the circuit shown in figure below.

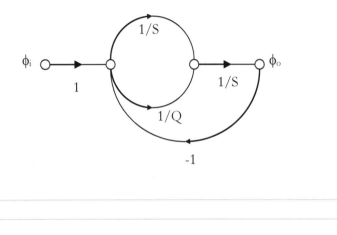

[18.3] Find out the output DCD for the normalized VLSI clock distribution circuit shown in figure below. Assuming the input clock has 50% duty cycle and the buffer is ideal and it has infinite gain and threshold voltage of $Vcc/2 + \delta V$.

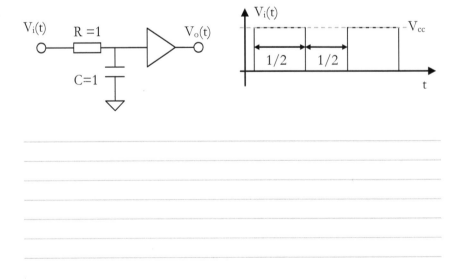

[18.4] Find out the output DCD for the normalized VLSI clock distribution circuit shown in figure below. Assuming the buffer is ideal and it has infinite gain and threshold voltage of $Vcc/2 + \delta V$.

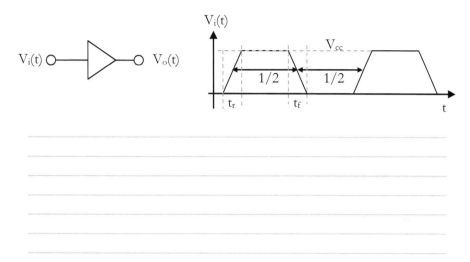

[18.5] Derive the input to output DCD transfer function for the normalized VLSI clock distribution circuit shown in figure below. Assuming the buffer is ideal and it has infinite gain and threshold voltage of $Vcc/2$.

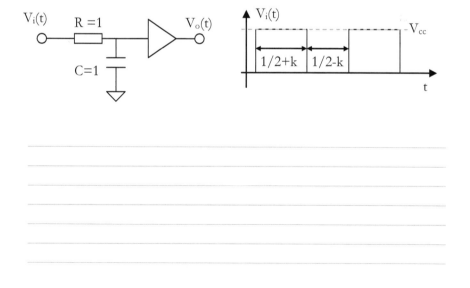

[18.6] Find the I/Q phase error transfer function for the 4-phase polyphase clock signal {ck1, ck2, ck3, ck4} passing through a polyphase filter circuit shown in figure below (assuming RC $=1/\omega$).

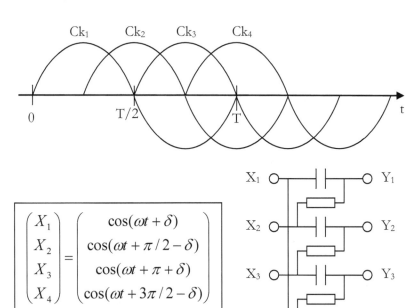

$$\begin{pmatrix} X_1 \\ X_2 \\ X_3 \\ X_4 \end{pmatrix} = \begin{pmatrix} \cos(\omega t + \delta) \\ \cos(\omega t + \pi/2 - \delta) \\ \cos(\omega t + \pi + \delta) \\ \cos(\omega t + 3\pi/2 - \delta) \end{pmatrix}$$

[18.7] Find the I/Q phase error transfer function for the 4-phase polyphase clock signal {ck1, ck2, ck3, ck4} passing through a polyphase filter circuit shown in figure below (assuming RC $=1/\omega$).

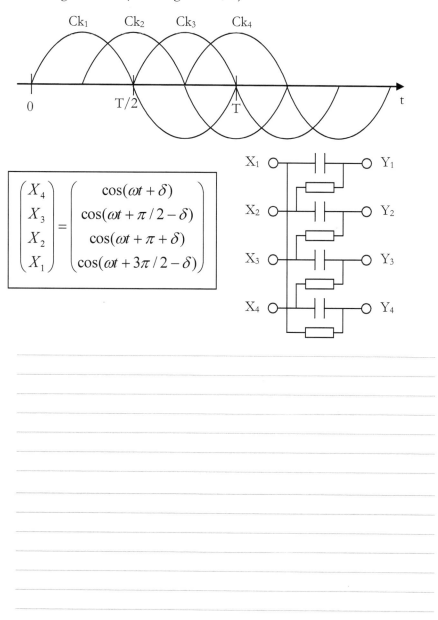

$$\begin{pmatrix} X_4 \\ X_3 \\ X_2 \\ X_1 \end{pmatrix} = \begin{pmatrix} \cos(\omega t + \delta) \\ \cos(\omega t + \pi/2 - \delta) \\ \cos(\omega t + \pi + \delta) \\ \cos(\omega t + 3\pi/2 - \delta) \end{pmatrix}$$

[18.8] Explain why the circuit shown below can be used to minimize the DCD in the clock distribution circuit.

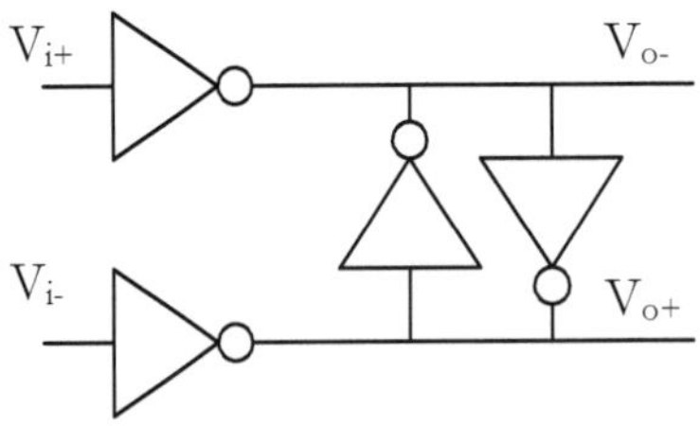

18.2 SAMPLE INTERVIEW QUESTIONS

1. What does it mean by jitter amplification effects?

2. Explain jitter amplification effect of a VLSI high-speed I/O channel.

3. Why AC coupling can usually be used to compensate for the DCD of a VLSI clock signal?

4.   Is DRC a phase domain or filter-like circuit?

5.   What is likely the jitter transfer function of a second order PLL circuit?

6.   Why filter can be viewed as a jitter filter?

7. Why PLL with first-order loop filter has a second order jitter transfer function characteristics?

8. What is likely the order of the DRC loop for high-speed I/O with first-order loop filter and a PI circuit?

9. What is a polyphase filter?

10. Why polyphase filter can be used to compensate for I/Q phase error of a high-speed I/O using 4-phase clock?

11. For a CMOS inverter based clock buffer chain, why the DCD is usually getting worse from stage to stage?

19

SIGNAL INTEGRITY AND POWER DELIVERY

- Ringing Effects in High-Speed I/O Circuits

- Delay Effects in High-Speed I/O Circuits

- Cross-Talk Effects in High-Speed I/O Circuits

- Power Delivery Induced Noises

VLSI high-speed I/O circuits are facing increasingly severe design challenges related to the signal integrity (SI) and power delivery (PD) in high-speed data transmission. Major signal integrity and power delivery issues in high-speed I/O circuits are related to ringing, reflection, delay variation (or jitter), crosstalk, dI/dt and IR noise effects.

These signal integrity and power delivery issues previously appeared only in the PCB circuit interconnects that are related to the non-ideal effects of the metal routing, such as the parasitic resistance effects at low frequency, the parasitic capacitance effects at midrange frequency and the parasitic inductance effects at high frequency. These issues are becoming increasingly problematic in the VLSI high-speed I/O circuits due mainly to the increasing data rate, clock frequencies, signal slew rate and interconnect complexity. Signal integrity and power delivery issues are now among major concerns in both high-speed I/O PCB and VLSI on-chip circuit design, that are main roots causes of pattern-dependent or soft failures, which are difficult to debug, resulted in either wrong data captured due to poor data signal quality or data captured at wrong time due to poor clock signal timing.

19.1 HOMEWORK AND PROJECT PROBLEMS

[19.1] The clock of a VLSI high-speed I/O circuit is given in figure below, where T = 200ps and tr = 20ps. What is likely the frequency range the clock power is located?

[19.2] A VLSI clock channel can be modeled using a LRC circuit shown in figure below. What is likely the maximum L value that can be used to pass the clock signal given in problem above?

[19.3] For the VLSI clock channel and the clock signal given in problem [19.1] and [19.2]. What is the preferred R for the best SI design requirement?

[19.4] If a VLSI high-speed I/O channel can be modeled using a RLC network as shown in figure below. Assume Rtx > $(L/C)^{0.5}$. Determine the rise time t_r such that there is no ringing occur at the output.

[19.5] If a VLSI high-speed I/O channel can be modeled using a RLC network as shown in figure below. Assume Rtx < $(L/C)^{0.5}$. Determine the rise time t_r such that there is no ringing occur at the output.

[19.6] If a VLSI high-speed I/O channel can be modeled using a RLC network as shown in figure below. Assume Rrx > $(L/C)^{0.5}$. Determine the rise time t_r such that there is no ringing occur at the output.

[19.7] If a VLSI high-speed I/O channel can be modeled using a RLC network as shown in figure below. Assume $R_{rx} < (L/C)^{0.5}$. Determine the rise time t_r such that there is no ringing occur at the output.

[19.8] If a VLSI high-speed I/O channel can be modeled using a RLC network as shown in figure below. Determine the rise time t_r such that there is no ringing occur at the output.

[19.9] If a VLSI high-speed I/O channel can be modeled using a RLC network as shown in figure below. At what condition (in terms of resistances, L and C) no ringing occur at the output at all rise time condition?

[19.10] A typical VLSI electronic system power delivery model is shown in figure below. Estimate the range of the component parameters in this circuit based on a real system (such as a CPU in a tablet PC).

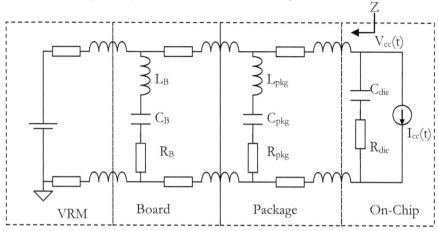

Decreasing in L and C values

[19.11] Identify the region of the impedance profile in the figure shown in below. Estimate the peaking frequency and DC values of the impedance profile for a tablet PC.

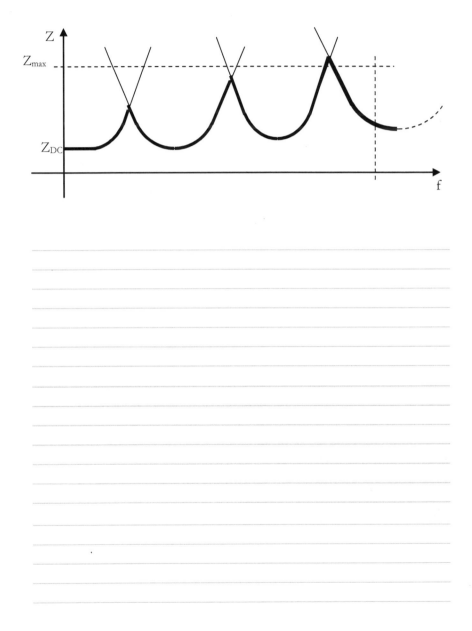

19.2 SAMPLE INTERVIEW QUESTIONS

1. What is crosstalk and how can it be avoided?

2. What is the power IR drop and di/dt noise? How to minimize the?

3. How will power supply line IR and di/dt noise impact the performance of the VLSI high-speed I/O circuits?

4. What is the ground bounce?

5. How will the ground bounce impact the performance of the VLSI high-speed I/O circuits?

6. What is the ringing effect?

7. How will the ringing effects impact the performance of the VLSI high-speed I/O circuits?

8. What are the reflection effects?

9. How will the reflection effect impact the performance of the VLSI high-speed I/O circuits?

10. What are the effects of the decoupling capacitors?

11. Why power supply decoupling is needed at various circuit levels at the VRM, the board, the package, and the chip levels?

12. What happens if metal wires are very near to each other and what happens if the metals overlap each other?

13. What are the measures a design can take for meeting signal-integrity targets?

14. Why is the pad ring provided with power supply connections that are separate from those of the core design?

15. In processes that have multiple layers of metal interconnect, why is it common to make the upper wires thicker than the lower layers?

16. Why is the power supply interconnect layout planned out before other elements?

17. Similarly, why are busses, differential signals, and shielded signals routed before other general signals?

18. What are the roots and resistance styles of power supply layout?

19. Why clock skew minimization in high-speed I/O is a major design challenge?

20. What are the major advantages and disadvantages of using a single clock tree conductor driven by one big buffer?

21. In ASIC design flows, why clock trees are inserted after the logic cells has been placed?

22. In clock tree design, how is clock skew minimized at the leaves of the tree?

23. What is a routing channel?

24. Why are routing channels used in IC layouts?

25. When routing a signal interconnect, why is it desirable to minimize layer changes through vias?

26. Interconnect resistance is usually minimized in IC layouts. Give at least four situations where a deliverable large, but controlled, resistance is usually required?

27. Why should minimum-width paths be avoided in the design of deliberate resistances?

28. Why we want to minimize the capacitance of electrical nodes in an IC design. Give four examples of circuits where one would wish a larger, but controlled, capacitance at a node?

29. Explain what is meant by electromigration. What are some possible consequences of unexpectedly high electromigration?

30. How is electro migration controlled in IC layout design?

31. Why do we use wide metal conductors as power grid?

32. When placing multiple vias to connect two metal conductors, why is it better to space the vias far apart from each other?

33. Why on-die R L C parameters are usually much smaller than the package and board RLC parameters in a typical power delivery system for a tablet PC?

34. What is the commonly used ratio S/H to minimize the cross-talk effects of the two signal route?

35. What is the ringing effect in VLSI high-speed I/O circuit? What is the root cause of the ringing effect?

36. What are the cross-talk effects? What are the causes of the cross-talk effect?

20

TDR & VNA MEASUREMENT TECHNIQUES

- TDR Measurement Basis
- TDR Electrical Signatures
- VNA Measurement Basis

The time-domain reflectometer (TDR) has been a standard measurement tool for characterizing and troubleshooting high-speed I/O circuits. The vector network analyzer (VNA), on the other hand, is becoming more commonly used signal integrity measurement tools as the result of increasing differential signaling speed and the need for more accurate characterization and modeling of differential interconnects (such as cables, connectors and printed circuit boards).

The TDR measurements are based on the time-domain stimulation of the devices under test (DUT) with voltage steps. The time delay and the signal magnitude of the reflection of the voltage step traveling through the DUT can be captured to extract circuit parameters such as the length and impedance information of the DUT.

On the other hand, the measurements with PNA are implemented in the frequency domain. where the DUT are characterized at each frequency of interest, one point at a time. The magnitude and phase shift from the DUTs are measured relative to the incident signals. The phase shifts are then related to the length of the DUTs, the longer the DUTs the larger the phase shifts, also, the higher the frequencies the larger the phase shifts.

20.1 HOMEWORK AND PROJECT PROBLEMS

The following problems are based on the TDR setup shown in figure below.

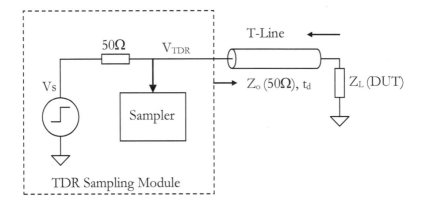

[20.1] Determine the impedance of the DUT based on the following TDR curve.

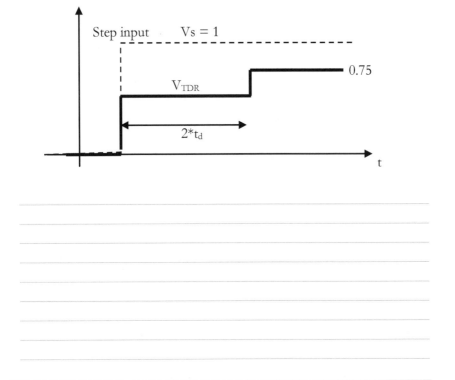

[20.2] Determine the impedance of the DUT based on the following TDR curve.

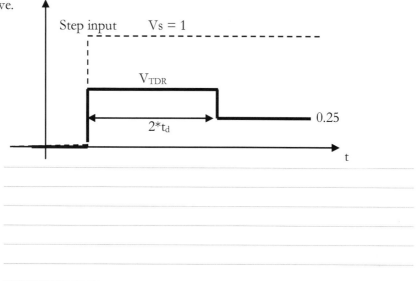

[20.3] Based on the following TDR curve, estimate type of device or device combination in the DUT. Determine the value of R in the DUT.

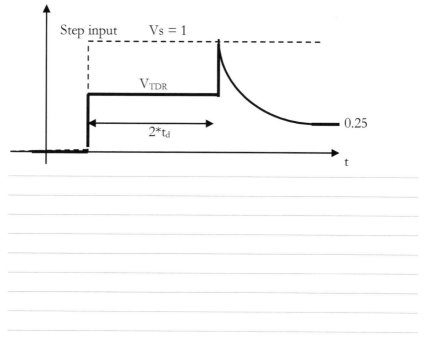

[2.4] In above problem, if td = 1ns, estimate the parameters of all devices in the DUT.

[20.5] Based on the following TDR curve, estimate type of device or device combination in the DUT. Determine the value of R in the DUT.

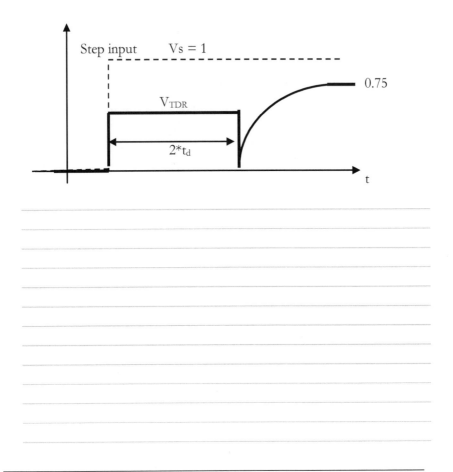

[20.6] In above problem, if td = 1ns, estimate the parameters of all devices in the DUT.

[20.7] Based on the following TDR curve, estimate type of device or device combination in the DUT. Determine the value of R in the DUT.

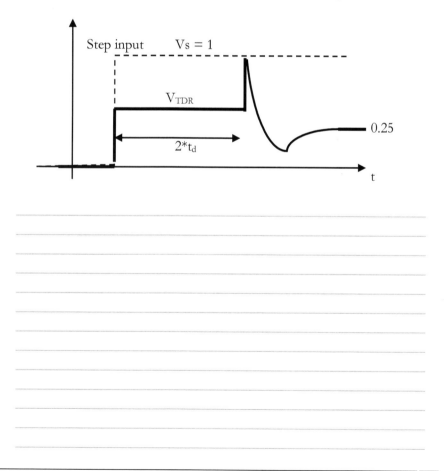

[20.8] Based on the following TDR curve, estimate type of device or device combination in the DUT. Determine the value of R in the DUT.

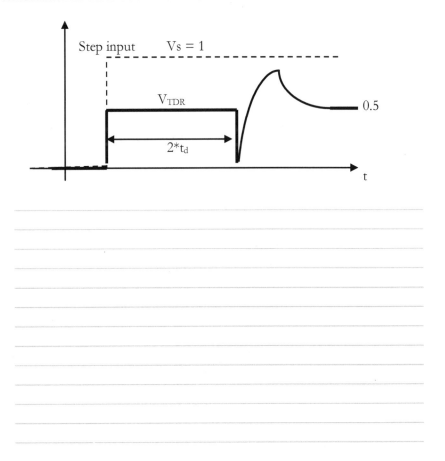

20.2 SAMPLE INTERVIEW QUESTIONS

1. Explain how a TDR works.

2. What principle does the time domain reflectometer use to test cables?

3. How will TDR be used in VLSI high-speed I/O development?

21

VLSI HIGH-SPEED I/O REFERENCE AND BIASING CIRCUITS

- Vt-Based Current Bias

- Vbe- Based Current Bias

- Constant Gm Current Bias

- VLSI Bandgap Voltage Reference Circuits

- VLSI PTAT Reference Circuits

Most analog circuits in VLSI high-speed I/O require certain voltage and current bias for the proper operations. On the other hand, circuit blocks such as Tx swing control, adaptive on-die termination, squelch circuit and LDO require voltage reference with absolute accuracy better than 5% that are independent of PVT conditions.

Several current bias such as constant-gm, PTAT, Vt-based, and Vbe-based current bias can be realized using VLSI circuits. On the other hand, PVT insensitive voltage reference can be generated using VLSI bandgap circuits.

21.1 HOMEWORK AND PROJECT PROBLEMS

[21.1] Derive the expression of V_{BE1} and V_{BE2} in the figure below in terms of I, N_1 and N_2.

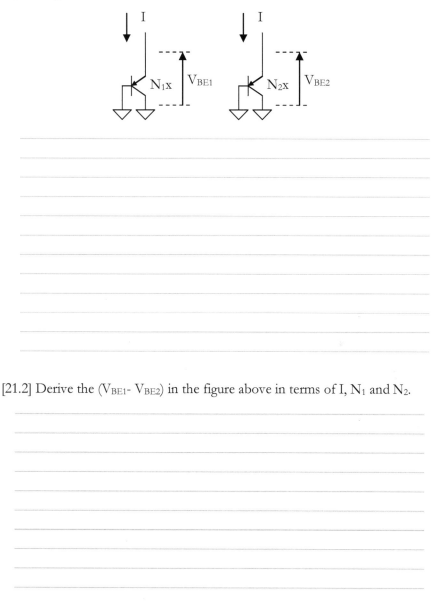

[21.2] Derive the (V_{BE1}- V_{BE2}) in the figure above in terms of I, N_1 and N_2.

[21.3] Plot of (V_{BE1}- V_{BE2}) in the figure above versus T. What is the slope and intercept?

[21.4] Assuming all NMOS and PMOS devices are identical respectively. Derive the current IB in the figure below in terms of V_T, β, and R.

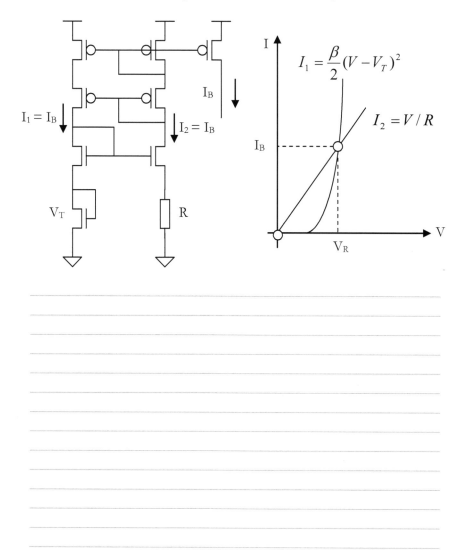

[21.5] Assuming all NMOS and PMOS devices are identical respectively. Derive the current IB in the figure below in terms of V_{BE}, and R.

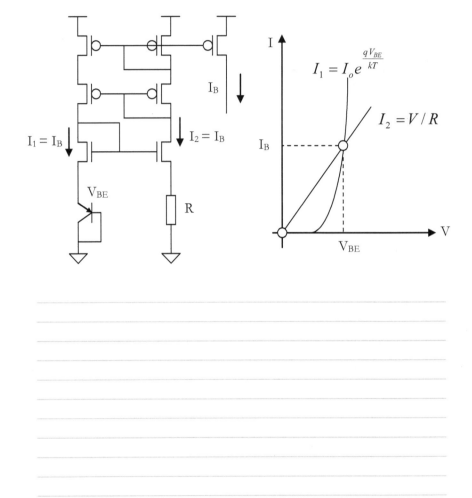

[21.6] Assuming all NMOS and PMOS devices are identical respectively. Derive the current IB in the figure below in terms of V_T, β, and R.

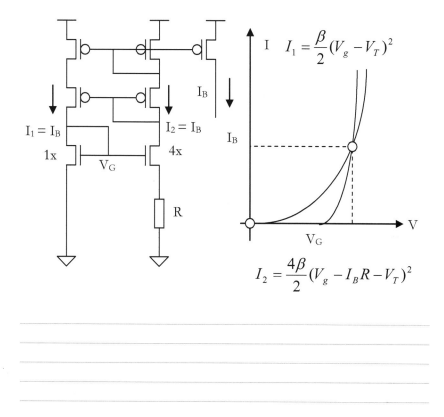

$$I_1 = \frac{\beta}{2}(V_g - V_T)^2$$

$$I_2 = \frac{4\beta}{2}(V_g - I_B R - V_T)^2$$

[21.7] Derive the voltage V_{BG} in the figure below in terms of N, R, and R1.

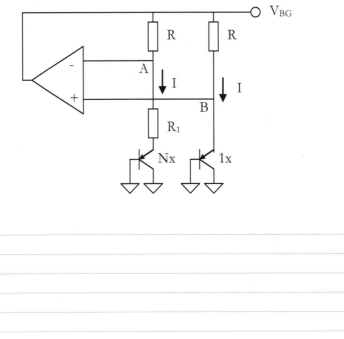

[21.8] For the bandgap circuit shown in above figure, what is minimal supply voltage for the circuit to work?

[21.9] Derive the voltage V_{BG} in the figure below in terms of N, R, and R1.

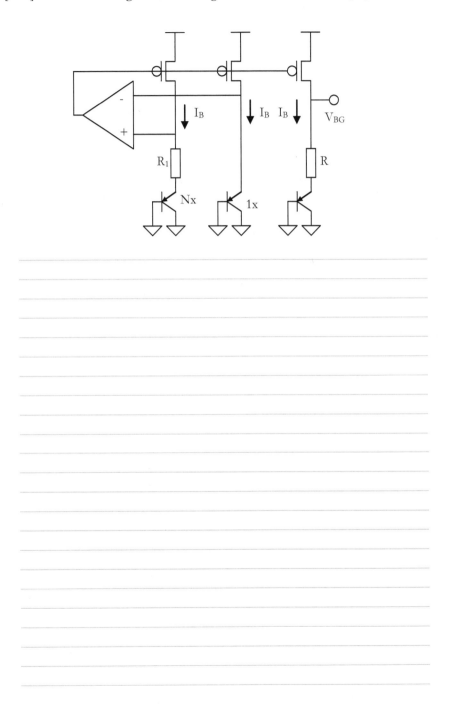

[21.10] For the bandgap circuit shown in above figure, what is minimal supply voltage for the circuit to work?

[21.11] Derive the voltage V_{BG} in the figure below in terms of N, R, R1, and R2.

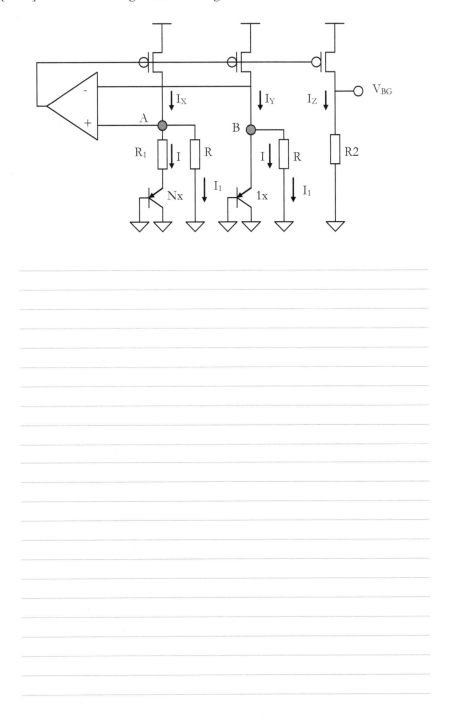

[21.12] For the bandgap circuit shown in above figure, what is minimal supply voltage for the circuit to work?

[21.13] Derive the voltage V_{BG} in the figure below in terms of N, R, and R1.

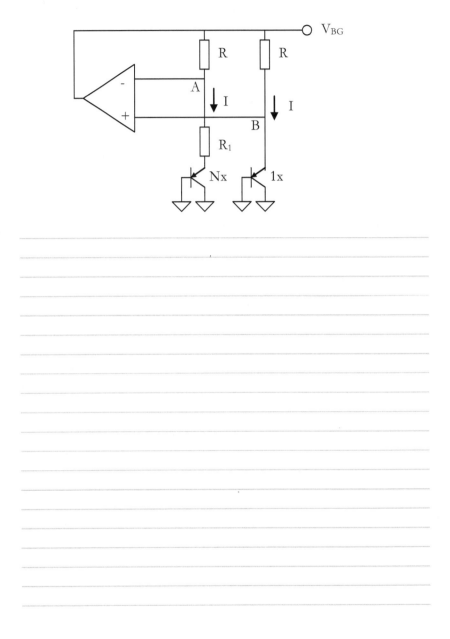

[21.14] For the bandgap circuit shown in above figure, what is minimal supply voltage for the circuit to work?

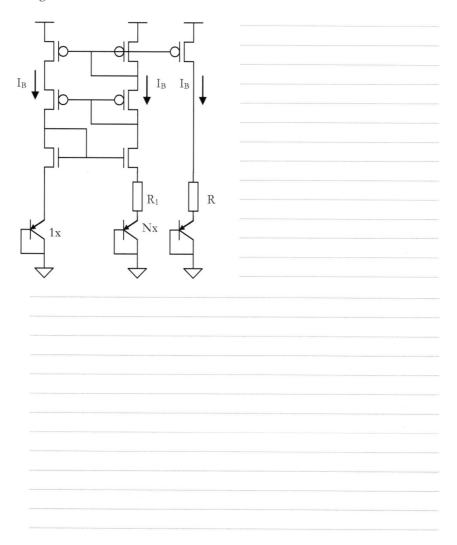

21.2 SAMPLE INTERVIEW QUESTIONS

1. How does a pn junction works?

2. Explain how a constant-Gm bias circuit works.

3. Explain how a Vt-based bias circuit works.

4. Explain how a PTAT reference works.

5. Explain how a bandgap reference circuit works.

6. Why on-chip current bias cannot be realized with good absolute accuracy?

7. Why bandgap voltage reference usually needs a start-up circuit?

8. Explain how the start-up circuit works.

APPENDIX I:
0.18um CMOS Technology File

AI.1 Passive Circuit Elements:

Parameters	N+	P+PLY	P+	POLY	M1	M2	M3	Units
Sheet-R	6	250	5.9	6.1	0.05	0.05	0.05	Ω/sq
Contact-R	7	7	7	7		0.7	1.3	Ω

Gate Oxide Thickness : 45   angstrom

Parameters	M4	M5	N+BLK	P+BLK	N_W	NS	POLY*	
Sheet-R	0.02	0.01	70	100	300	8	1500	Ω/sq
Contact-R	1.6	2.0						Ω

(POLY*: silicide blocked)

Capacitance	N+	P+	POLY	D_N_W	N_W	
Area (substrate)	900	1200			170	aF/um^2
Area (N+active)			8000			aF/um^2
Area (P+active)			8000			aF/um^2
Area (r well)	24					aF/um^2
Area (NMOS Cap)			2400			aF/um^2
Area (NMOS varactor)			6000			aF/um^2
Area (N+ varactor)		900				aF/um^2
Fringe (substrate)	100	50				aF/um
Overlap (N+active)			400			aF/um
Overlap (P+active)			400			aF/um

AI.2 MOS Device Models

.MODEL NMOS NMOS (LEVEL = 49

+VERSION = 3.1   TNOM   = 27   TOX   = 4.4E-9

+XJ   = 1E-7   NCH   = 2.3549E17   VTH0   = 0.3145

+K1   = 0.502334   K2   = -0.010685   K3   = 1E-3

+K3B   = 5.9241   W0   = 1E-7   NLX   = 3.75879E-7

+DVT0W = 0   DVT1W = 0   DVT2W = 0

+DVT0   = 0.44187   DVT1   = 0.09826   DVT2   = -0.5

+U0   = 293.104   UA   = -1.304E-9   UB   = 2.567E-18

+UC   = 5.74371E-11   VSAT   = 1.5653E5   A0   = 1.2236

+AGS   = 0.3065   B0   = 1.18665E-6   B1   = 5E-6

+KETA   = 0.02136   A1   = 0   A2   = 0.5787

+RDSW   = 150   PRWG   = 0.19046   PRWB   = -0.2

+WR   = 1   WINT   = 6.3905E-9   LINT   = 1.115E-8

+DWG   = 1.356E-9   DWB   = 2.018E-8   VOFF   = -0.09375

+NFACTOR = 2.1751   CIT   = 0   CDSC   = 2.4E-4

+CDSCD   = 0   CDSCB   = 0   ETA0   = 1.622E-3

+ETAB   = 3.764E-4   DSUB   = 0   PCLM   = 1.1109

+PDIBLC1 = 1   PDIBLC2 = -1.61799E-3   PDIBLCB = -0.1

+DROUT = 0.9149   PSCBE1 = 9.129E9   PSCBE2 = 7.0002E-10

+PVAG   = 0.09162   DELTA   = 0.01   RSH   = 6.3

+MOBMOD = 1   PRT   = 0   UTE   = -1.5

+KT1   = -0.11   KT1L   = 0   KT2   = 0.022

+UA1   = 4.31E-9   UB1   = -7.61E-18   UC1   = -5.6E-11

+AT   = 3.3E4   WL   = 0   WLN   = 1

+WW   = 0   WWN   = 1   WWL   = 0

+LL   = 0   LLN   = 1   LW   = 0

+LWN   = 1   LWL   = 0   CAPMOD = 2

+XPART   = 0.5   CGDO   = 3.59E-10   CGSO   = 3.59E-10

+CGBO   = 1E-9   CJ   = 8.321E-4   PB   = 0.826

+MJ   = 0.5339   CJSW   = 1.295E-10   PBSW   = 0.8

+MJSW   = 0.3433   CJSWG   = 3.3E-10   PBSWG   = 0.8

+MJSWG   = 0.3433   CF   = 0   PVTH0   = -9.89E-3

+PRDSW   = 1.58446   PK2   = 4.191E-3   WKETA   = -2.7151E-3

+LKETA = -4.0072E-3 PU0   = -4.417625   PUA   = -4.97062E-11

+PUB   = 2.421E-23 PVSAT = 86.81   PETA0 = -1E-4

+PKETA = -4.6996E-3)

.MODEL PMOS PMOS (LEVEL = 49

+VERSION = 3.1   TNOM = 27   TOX   = 4.4E-9

+XJ   = 1E-7   NCH   = 4.159E17   VTH0 = -0.4011

+K1   = 0.6031   K2   = -2.0191E-3   K3   = 0.0931728

+K3B = 19.918   W0   = 1E-6   NLX   = 4.2082E-8

+DVT0W = 0   DVT1W = 0   DVT2W = 0

+DVT0 = 0.73003   DVT1 = 0.6076   DVT2 = -0.3

+U0   = 121.184   UA   = 1.6516E-9   UB   = 2.0102E-21

+UC   = -1E-10   VSAT = 1.509E5   A0   = 1.054245

+AGS = 0.27903   B0   = 1.3535E-6   B1   = 5E-6

+KETA = 0.0151   A1   = 4.077091E-3   A2   = 1

+RDSW = 653.549   PRWG = -7.9077E-4 PRWB = -0.3911617

+WR   = 1   WINT = 0   LINT = 2.9635E-8

+DWG = -3.567E-8 DWB   = -8.697E-9 VOFF = -0.13832

+NFACTOR = 1.29143   CIT   = 0   CDSC = 2.4E-4

+CDSCD = 0   CDSCB = 0   ETA0 = 1.077418E-3

+ETAB = -2.263E-3   DSUB = 3.16648E-3 PCLM = 0.51831

+PDIBLC1 = 0.05821   PDIBLC2 = 0.01809   PDIBLCB = -1E-3

+DROUT = 0.956   PSCBE1 = 1.649E9   PSCBE2 = 5E-10

+PVAG = 0.01499   DELTA = 0.01   RSH   = 5.9

+MOBMOD = 1   PRT   = 0   UTE   = -1.5

+KT1   = -0.11   KT1L = 0   KT2   = 0.022

+UA1   = 4.31E-9   UB1   = -7.61E-18   UC1   = -5.6E-11

+AT   = 3.3E4   WL   = 0   WLN   = 1

+WW   = 0   WWN   = 1   WWL   = 0

+LL   = 0   LLN   = 1   LW   = 0

+LWN   = 1   LWL   = 0   CAPMOD = 2

+XPART = 0.5   CGDO = 3.52E-10   CGSO = 3.52E-10

+CGBO = 1E-9   CJ   = 1.2042E-3 PB   = 0.88235

+MJ   = 0.44104   CJSW = 1.79054E-10 PBSW = 0.8

+MJSW   = 0.2659   CJSWG   = 4.22E-10   PBSWG = 0.8

+MJSWG = 0.265   CF   = 0   PVTH0 = 1.2495E-3

+PRDSW = -5   PK2   = 3.643765E-4   WKETA = 0.03172

+LKETA = -0.01104 PU0   = 2.22102   PUA   = 1.363E-10

+PUB   = 0   PVSAT = 50   PETA0 = 1E-4

+PKETA = -3.40506E-3   )

APPENDIX II:

Past EEE598 Exam Sheets

Midterm Exam 1

EEE598 VLSI High-Speed I/O Circuits

Fall 2008
Arizona State University
Instructor: Dr. Hongjiang Song
Exam Time: 6:30pm – 7:45pm Wednesday, Oct. 8, 2008

Print Name: _____

Signature: _____

ID#: _____

General instruction: This is a closed book/notes exam. However, you may bring a piece of paper (8 x11) with useful notes. You may also bring a calculator to exam.

Hint:

Good luck!

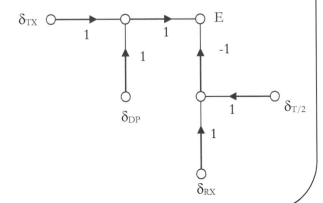

Problem 1 A VLSI high-speed I/O circuit structure is shown in Fig. 1.1. For simplicity we assume $t_{setup} = t_{hold} = 0$ for the Rx in the following questions and T is the clock period.

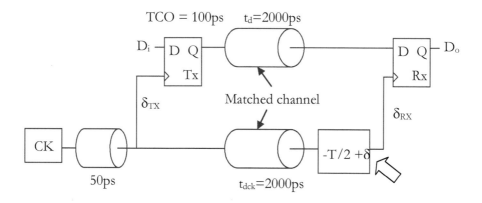

Fig.1.1 Forward clock I/O circuit

1.1) (5 points) For optimal circuit operation (i.e. maximum timing margin), what should be the value of δ in the circuit?

1.2) (10 points) Assume that the circuit was designed and optimized at typical corner with all parameters **designed at this corner,** if the delay of the two channels may vary by about 20% in PCB manufacture (i.e. $t_d - t_{dck} = +/-$ 400ps), what is the <u>maximum data rate</u> (or clock frequency) such a I/O circuit can be operated at? (Please show your work).

1.3) (5 points) For problem in (1.2) if we use a <u>delay locked loop</u> (you don't need to know its operation to answer this question) that adaptively controls δ for the delay uncertainty compensation, what should be the <u>minimal adjust range</u> of δ for completely compensating for the about PCB manufacture induced delay uncertainty?

1.4) (5 points) For the design problem in (1.3) if δ is adaptively controlled by a delay locked loop with a control accuracy (or delay uncertainty) of +/-50ps, what is the maximum I/O data rate (or clock frequency) if other delay uncertainty effects are ignored?

Problem 2.1 (9 points): For the following T-line circuits shown below. Please draw the corresponding voltage signals waveforms at point A, B, and C separately on the graphs below.

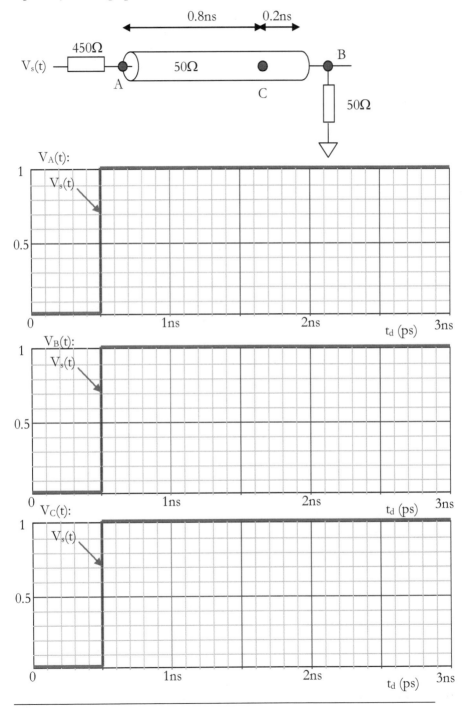

Problem 2.2 (9 points): For the following T-line circuits shown below. Please draw the corresponding voltage signals waveforms at point A, B, and C separately on the graphs below.

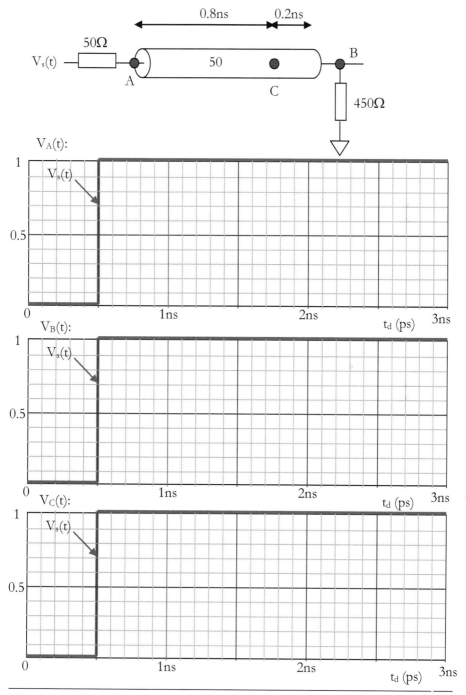

Problem 3. (Quick questions). Please **circle** the best answer to each question (only **ONE** selection per question please!). **Don't spend too much time on an individual question**. It is OK to guess the answer if you are not very sure about the correct answer to that question.

(3.1-3.4) Delay time of a VLSI buffer versus the load capacitance (in terms of fanout or FO) is shown in Figure below. Please answer the following questions.

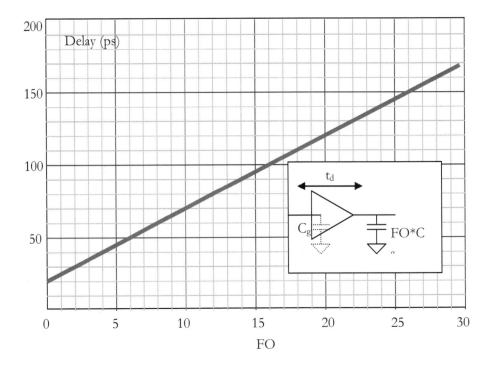

3.1) (5 points) What is likely the delay of the circuit with the load of FO = 36?

A) 50s   B) 100ps   C) 150ps   D) 200ps

3.2) (5 points) For the FO = 36 case, what is likely the delay of the circuit if a buffer is inserted between the load and the driver? (Assuming that all buffer stages have the same fanout value)

A) 50s   B) 100ps   C) 150ps   D) 200ps

3.3) (5 points) For the FO = 36 case, what is likely the delay of the circuit if <u>two buffers</u> are inserted between the load and the driver? (Assuming that all buffer stages have the same fanout value). (Hint: $36^{1/3} = 3.3$)

A) 6ns   B) 11ns   C) 16ns   D) 21ns

3.4) (5 points) About how many buffers should be inserted such that the delay of above circuit is minimized? (Hint: $36^{1/4} = 2.45$, $36^{1/5} = 2.05$, $36^{1/6} = 1.82$)

A) 1   B) 2   C) 3   D) 4

(3.5-3.7) Delay time of a buffered VLSI interconnect (with identical load as driver) versus the interconnect length is shown in Figure below. Please answer the following questions.

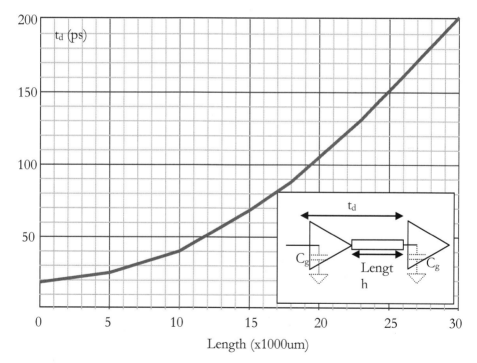

3.5) (5 points) What is likely the new delay t_d of the 30,000um interconnect path if <u>a repeater</u> of same size as original buffer is inserted?

A) 100ps   B) 120ps   C) 140ps   D) 160ps

3.6) (5 points) What is likely the delay t_d of the 30,000um interconnect path if <u>two repeaters</u> of same size as original buffer are inserted?

A) 100ps   B) 120ps   C) 140ps   D) 160ps

3.7) (5 points) <u>How many repeaters</u> should be inserted to provide the minimal delay t_d of above 30,000um interconnect path?

A) 2   B) 3   C) 4   D) 5

3.8) (5 points) The setup and hold times of the receiver are important timing parameters of the high-speed I/O circuits. Which of the following cases is <u>not likely a true result</u> of the receiver setup and hold time simulation?

(i) $t_{setup} = 20ps$, $t_{Hold} = 100ps$;   (ii) $t_{setup} = -20ps$, $t_{Hold} = 100ps$;

(iii) $t_{setup} = 20ps$, $t_{Hold} = -100ps$;   (iv) $t_{setup} = 0ps$, $t_{Hold} = 100ps$;

A) (ii)   B) (ii) and (iii)   C) (iii)   D) (iv)

(3.9-3.10) Ringing effect is one common source of jitters in the high-speed I/O. Ringing effect in the following circuit can be eliminated by adjustment of the circuit components. (Ignore the output capacitance of the gate for the following questions)

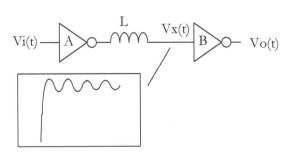

3.9) (5 points) The ringing effect can be eliminated by reducing the size of the device A to half, (i.e. 0.5x) what is likely the size of device B that can also be used to eliminate the ringing effect?

A) 0.25x   B) 0.5x   C) 2x   d) 4x

3.10) (5 points)   (5 points) if the ringing effect can be eliminated by reducing the size of the device A to half, what is likely the inductance L that can also be used to eliminating the ringing effect?

A) 0.25x   B) 0.5x   C) 2x   d) 4x

3.11) (5 points) For the circuit shown in figure below, which of the following recommendation is generally not true for minimizing the cross-talk induced jitter at point Vc

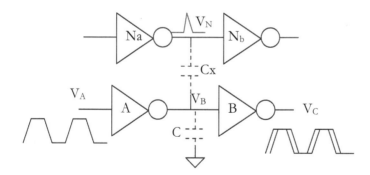

A) Reduce C_x;   B) Reduce size of device N_b;

C) Increase size of device A;   D) Reduce size of device Na.

3.13 (2 points) For the T-line circuit and signal waveform shown below, which is most likely the <u>true</u> case?

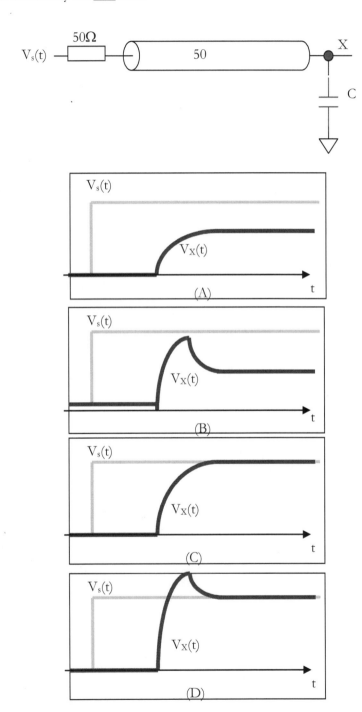

Midterm Exam 2

EEE598 VLSI High-Speed I/O Circuits

Fall 2008
Arizona State University
Instructor: Dr. Hongjiang Song
Exam Time: 6:30pm – 7:45pm Monday, Nov. 10, 2008

Print Name: _____

Signature: _____

ID#: _____

General instruction: This is a closed book/notes exam. However, you may bring a piece of paper (8 x11) with useful notes. You may also bring a calculator to exam.

Good luck!

Problem 1 A VLSI programmable (i.e. multi-mode) frequency divider circuit with state output [Q1, Q2] is shown in figure below. The operation modes of the circuit are shown in table.

Mode	Sx	Sy	Sz
X	Short	open	open
Y	open	Short	open
Z	open	open	Short

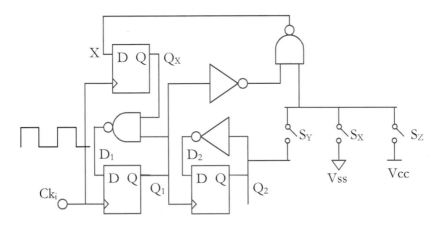

1.1) (15 points) What is the division ratio for <u>operation mode X</u>? (Please show your work)

Q1	Q2	Qx	D1	D2	X

1.2) (15 points) What is the division ratio for <u>operation mode Y</u>? (Please show your work)

Q1	Q2	Qx	D1	D2	X

1.3) (10 points) What is the division ratio for <u>operation mode Z</u>? (Please show your work)

Q1	Q2	Qx	D1	D2	X

Problem 2.1 (15 points): Shown below is a VLSI phase frequency detector (PFD) that is commonly used in VLSI PLL circuits. Draw the output waveform Up and Dn and R based on the input waveforms and the initial condition given below. (For simplicity, ignore the delay parameters (TCO, setup time, hold time etc.) of D-FF circuit and assume the delay of the <u>AND gate is 100ps</u>)

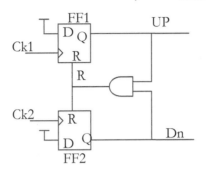

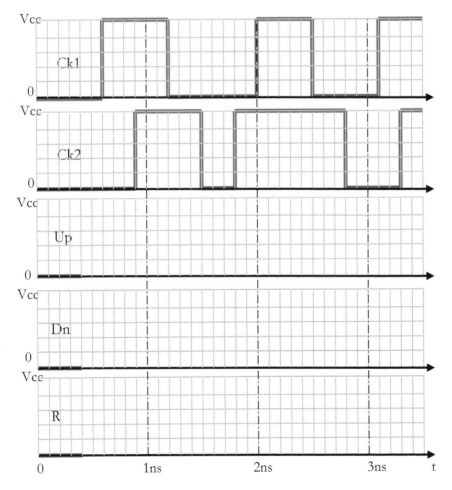

Problem 2.2 (15 points): Shown below is a VLSI phase frequency detector (PFD) that is commonly used in VLSI PLL circuits. Draw the output waveform Up and Dn and R based on the input waveforms and the initial condition given below. (For simplicity, ignore the delay parameters of D-FF circuit and assume the delay of the <u>AND gate is 100ps</u>)

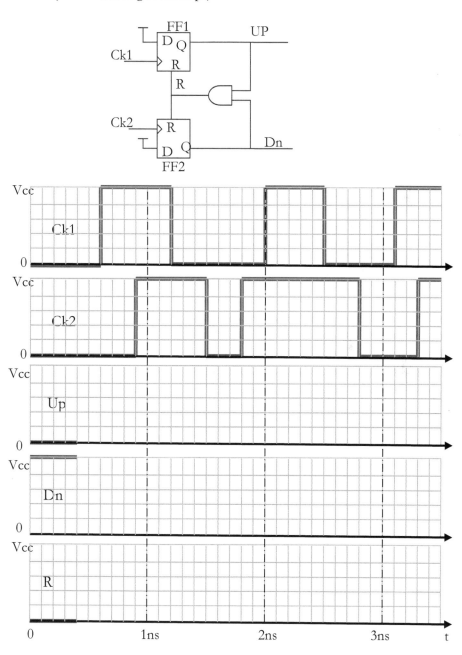

Problem 3. (Quick questions). Please **circle** the best answer to each question (only **ONE** selection per question please!). **Don't spend too much time on an individual question**. It is OK to guess the answer if you are not very sure about the correct answer to that question.

(3.1-3.2) A VLSI PLL circuit is shown below. Please answer the following questions:

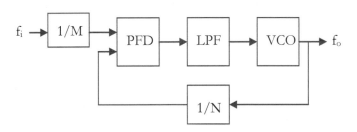

(3.1) (5 points) What is likely the output frequency of the PLL if the PLL circuit is properly locked-in.

a) $f_o = (NM)f_i$ b) $f_o = (N/M)f_i$

c) $f_o = (M/N)f_i$ d) $f_o = (1/(NM))f_i$

(3.2) (5 points, tricky one!) Suppose that there is a design problem in the divider circuit such that the feedback <u>divider may randomly (quickly change in-time) have two dividing ratio at very high frequency</u>: N (with the probability of p) and N1 (N1>>N) (with probability of 1-p). What PLL output frequency value(s) might be observed if a large number of manufactured PLL circuits are tested (assume M =1 for this question)?

a) $f_o = [pN+(1-p)N1]f_i$ b) $f_o = pNf_i$ or $(1-p)N1 \ f_i$

c) $f_o = Nf_i$ or $[pN+(1-p)N1]fi$ d) $f_o = [pN1+(1-p)N]f_i$

(3.3-3.4) The SFG of a 2nd order VLSI PLL circuit is shown below. Please answer the following questions:

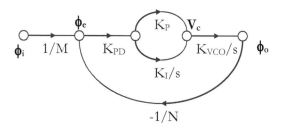

(3.3) (5 points) The bandwidth ω_n of the PLL is given as:

(a) $\sqrt{K_{PD}K_PK_{VCO}/N}$ (b) $\sqrt{K_{PD}K_PK_{VCO}/M}$

(c) $\sqrt{K_{PD}K_IK_{VCO}/N}$ (d) $\sqrt{K_{PD}K_IK_{VCO}/M}$

(3.4) (5 points) the damping factor $\zeta = 1/(2Q)$ of the PLL is given as:

(a) $\dfrac{\sqrt{K_{PD}K_PK_{VCO}/N}}{2K_I}$ (b) $\dfrac{\sqrt{K_{PD}K_IK_{VCO}/M}}{2K_P}$

(c) $\dfrac{\sqrt{K_{PD}K_PK_{VCO}/N}}{2\sqrt{K_I}}$ (d) $\dfrac{\sqrt{K_{PD}K_P^2K_{VCO}/N}}{2\sqrt{K_I}}$

(3.5) (5points) For the VLSI DLL circuit shown below, what is likely the <u>delay from A to B</u> after DLL has been properly locked-in.

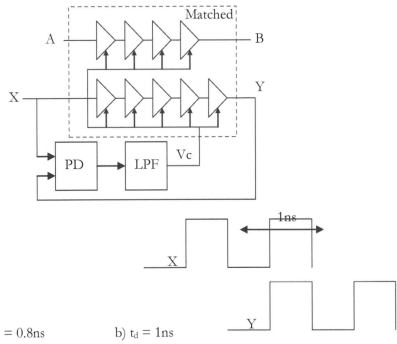

a) $t_d = 0.8$ns

b) $t_d = 1$ns

c) $t_d = 1.2$ns

d) $t_d = 1.25$ns

(3.6) (5 points) For the following VLSI phase interpolator circuit employing identical inverters, what is likely the output clock phase ϕ_x?

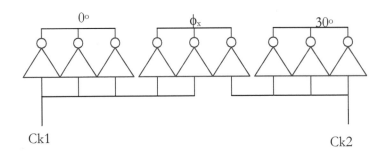

a) 10°

b) 15°

c) 20°

d) 25°

Final Exam

EEE598 VLSI High-Speed I/O Circuits

Fall 2008
Arizona State University
Instructor: Dr. Hongjiang Song
Exam Time: 4:50pm – 6:40pm Wednesday, Dec. 17, 2008

Print Name: _____

Signature: _____

ID#: _____

General instruction: This is a closed book/notes exam. However, you may bring a piece of paper (8 x11) with useful notes. You may also bring a calculator to exam.

Good luck!

Problem1. Shown in Fig.1.1 is a (simplified) VLSI high-speed I/O circuit. Please <u>ignore all gate delays and TCOs for the following questions</u>.

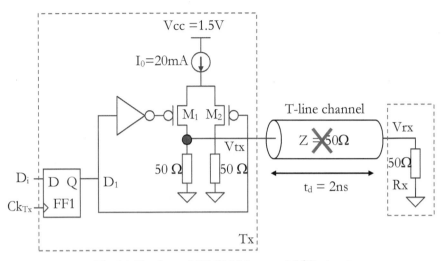

Fig.1.1 Single-end VLSI high-speed I/O circuit

1.1) 10 points) Plot the transmitter output **Vtx** for the given inputs <u>when the transmission line (channel) is **removed** from the circuit (i.e. open circuit)</u>:

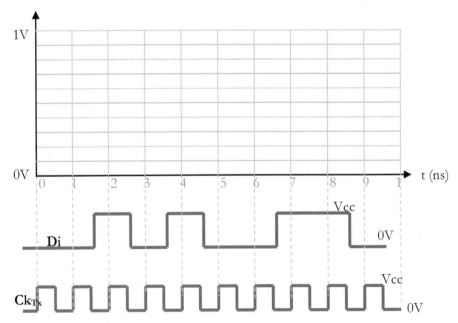

1.2) (10 points) Plot the transmitter output <u>**Vtx**</u> and receiver input <u>**Vrx**</u> <u>when</u> <u>the channel and Rx are connected (as shown in Fig.1.1):</u>

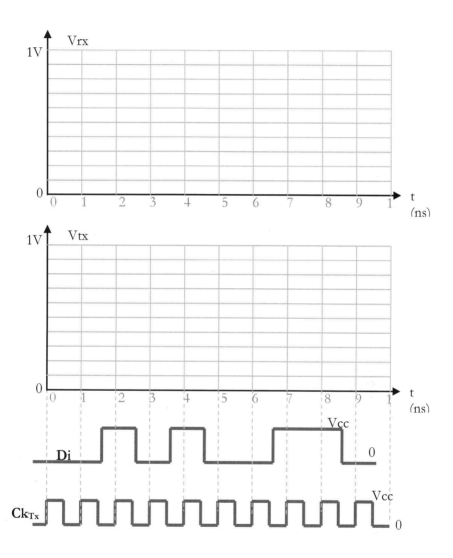

Problem 2. For the VLSI high-speed I/O data transmission circuit shown, solve the following problems (only one selection per question for (2.1) and (2.2)!):

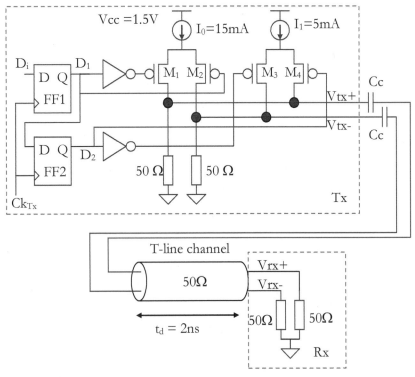

Fig.2 VLSI high-speed I/O differential data transmission circuit

2.1) (3 points) Device M3 and M4 are added in circuit for the <u>main</u> purposes of

(a) Channel Impedance termination   (b) dI/dt Noise Control

(c) Channel Equalization   (d) Slew Rate control

2.2) (3 points) The z-domain transmitter model is approximately given as ($|K|$ is a constant):

(a) $\dfrac{V_o(z)}{D_i(z)} \equiv \dfrac{V_{o+}(z) - V_{o-}(z)}{D_i(z)} = K(\dfrac{3}{4} + \dfrac{1}{4}z^{-1})$

(b) $\dfrac{V_o(z)}{D_i(z)} \equiv \dfrac{V_{o+}(z) - V_{o-}(z)}{D_i(z)} = K(\dfrac{3}{4} - \dfrac{1}{4}z^{-1})$

(c) $\dfrac{V_o(z)}{D_i(z)} \equiv \dfrac{V_{o+}(z) - V_{o-}(z)}{D_i(z)} = K(1 + \dfrac{1}{4}z^{-1})$

(d) $\dfrac{V_o(z)}{D_i(z)} \equiv \dfrac{V_{o+}(z) - V_{o-}(z)}{D_i(z)} = K(1 - \dfrac{1}{4}z^{-1})$

2.3) (10 points) (Tricky one!) <u>Ignore all gate delays and TCOs</u>. Plot the Transmitter output Vtx+ and Vtx- signal waveforms in the diagram below for the given inputs. (Assume Cc is very large and the steady state has been reached)

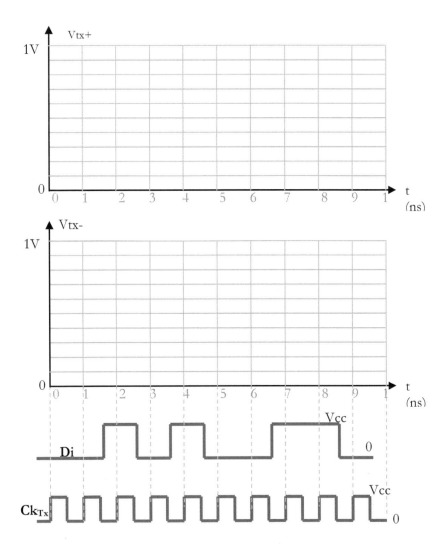

Problem 3. (Quick questions). Please **circle** the best answer to each question (only **ONE** selection per question please!). **Don't spend too much time on an individual question**. It is OK to guess the answer.

3.1 – 3.4) For the SFG and Rx jitter tolerance curve of a 5Gb/s high-speed I/O DRC circuit shown below, answer the following questions:

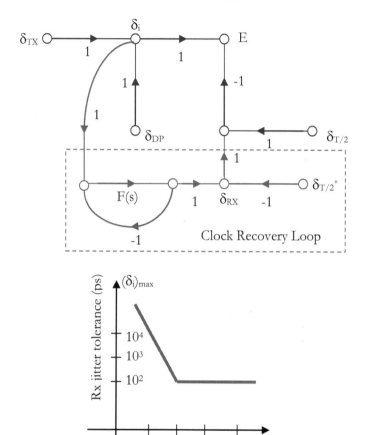

3.1) (3 points) The transfer function F(s) in generally has a _____ frequency response.

a) High-pass

b) Low-pass

c) Band-pass

d) Band-stop

3.2) (3 points) (Slightly tricky one!) Which of the following F(s) in likely used in above DRC loop?

(a) $F(s) = \dfrac{s}{\omega_o}$

(b) $F(s) = \dfrac{s}{\omega_o}\left(\dfrac{s}{\omega_o} + \dfrac{1}{Q}\right)$

(c) $F(s) = \dfrac{\omega_o}{s}$

(d) $F(s) = \dfrac{\omega_o}{s}\left(\dfrac{\omega_o}{s} + \dfrac{1}{Q}\right)$

3.3) (3 points) What is likely the bandwidth of the above DRC loop?

(a) $\omega_o \sim 10^5$ rad/s

(b) $\omega_o \sim 10^6$ rad/s

(c) $\omega_o \sim 10^7$ rad/s

(d) $\omega_o \sim 10^8$ rad/s

3.4) (3 points) If the 4-phase PLL you designed is used in this DRC loop for generating the receiver clock with a phase spacing error of 10ps (instead of 2ps as design spec.), the actual jitter tolerance curve will be shifted _____ compared with the curve based on the spec.

a) up

b) down

c) left

d) right

3.5-3.8) For the Bath Tub curve of a 5Gb/s high-speed I/O circuit shown below, answer the following questions

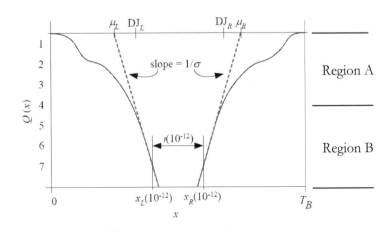

3.5) (3 points) The Bath Tub curve is likely related to

(a) The relationship between the eye opening of an I/O versus the channel length

(b) The relationship between the eye opening of an I/O versus its fanout

(c) The relationship between the eye opening of an I/O versus its data rate

(d) The relationship between the eye-opening of an I/O link versus the bit error rate

3.6) (3 points) What is likely the T_B value in above Bath Tub curve?

(a) 100ps   (b) 200ps

(c) 400ps   (d) 500ps

3.7) (3 points) <u>Region A</u> of the above Bath Tub curve is likely **not** dominated by

(a) deterministic jitter   (b) random Jitter

(c) dI/dt noise   (d) channel Impedance mismatch

3.8) (3 points) <u>Region B</u> of the above Bath Tub curve is likely dominated by

(a) deterministic jitter   (b) random Jitter

(c) dI/dt noise   (d) channel Impedance mismatch

3.9) (3 points) Which of the following is generally **not** a source to the ISI effect?

(a) dielectric loss   (b) skin effect

(c) radiation effect   (d) dI/dt noise

3.10) (3 points) Which of the following is likely **not** related to the ISI effect?

(a) channel length modulation effect   (b) bandwidth limitation

(c) DDJ   (d) signal filtering

3.11) (3 points) Which of the following circuit techniques is likely **not** related to channel equalization?

(a) FFE   (b) DEF

(c) Analog high-pass filter   (d) analog low-pass filter

3.12) (3 points) What is <u>DFE</u> in high-speed I/O circuits mean?

(a) defined filter equalization   (b) decision feedback equalization

(c) deterministic filter equalization   (d) digital filter equalization

3.13) (3 points) Where the <u>DFE</u> is usually used in high-speed I/O circuits?

(a) Tx clock generation circuit   (b) Rx clock generation circuit

(c) Tx circuit   (d) Rx circuit

3.14) (3 points) VLSI <u>Equalization circuit</u> techniques are likely used for solving the _____ issues :

(a) Reflection/ringing effect   (b) Cross-talk effects

(c) ISI effect   (d) Delay skew effect

3.15) (3 points) VLSI <u>on-die termination</u> circuit techniques are likely used for solving the _____ issues :

(a) Reflection/ringing effect   (b) Cross-talk effects

(c) ISI effect   (d) Delay skew effect

3.16) (3 points) VLSI <u>source synchronization</u> circuit techniques are likely used for solving the _____ issues :

(a) Reflection/ringing effect   (b) Cross-talk effects

(c) ISI effect   (d) Delay skew effect

3.17) (3 points) VLSI <u>differential circuit</u> techniques are likely used for solving the _____ issues :

(a) Reflection/ringing effect   (b) Cross-talk effects

(c) ISI effect   (d) Delay skew effect

3.18) (3 points) VLSI <u>LVDS circuit</u> techniques are likely **not** used for solving the _____ issues

(a) high-speed data transmission;   (b) low power data transmission;

(b) reduced cross-talk noise   (d) low device counter data transmission

3.19) (3 points) 3b/4b or 8/10b coding in high-speed I/O circuits are likely used for _____ :

(a) error correction;   (b) reducing power dissipation;

(b) ensuring DRC circuit operation   (d) reducing cross-talk

3.20) (3 points) the main reason the PLL <u>PFD</u> (phase frequency detector) is not suitable for data recover circuit:

(a) It is not fast enough

(b) it cannot be integrated

(c) It consumes too much power

(d) It cannot handle random data

3.21) (3 points) What are the other names of the <u>common-clock</u>, <u>forward-clock</u>, and <u>embedded-clock</u> I/O circuits?

(a) Synchronous, source synchronous, and data recovery I/O circuits

(b) Source synchronous, synchronous, and data recovery I/O circuits

(c) Synchronous, data recovery, and source synchronous I/O circuits

(d) Data recovery, Synchronous, source synchronization, and I/O circuits

3.22) (1 points) How many types of high-speed I/O circuits do you think are being used in this class room?

(a) 2

(b) 3

(c) 4

(d) 5

Midterm Exam 1

EEE598 VLSI High-Speed I/O Circuits
Fall 2010
Arizona State University
Instructor: Dr. Hongjiang Song
Exam Time: 6:00pm – 7:15pm Thursday, Sept. 23, 2010

Print Name: _____

Signature: _____

ID#: _____

General instruction: This is a closed book/notes exam. However, you may bring a piece of paper (8 x11) with useful notes. You may also bring a calculator to exam.

Hint:

Good luck!

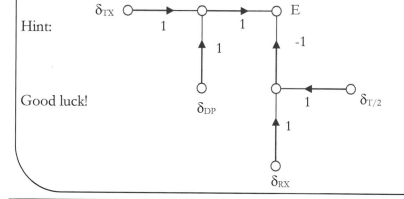

Problem 1 (Prototype I/O circuit model, I/O timing equation, VLSI circuit delay models) A VLSI high-speed I/O circuit structure is shown in Fig. 1.1. Such I/O circuit employs a <u>RC interconnect</u> based clock distribution circuit and an adjustable capacitor C for timing optimization at data rate of 500Mbps (i.e. T = 2ns). For simplicity, please ignore the setup/hold times of the Rx and the input parasitic capacitances of both Tx and Rx circuits.

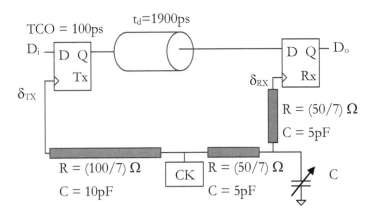

Fig.1.1 Common clock I/O circuit

1.1)   (10 points) For optimal circuit operation (i.e. maximum timing margin), what should be the value for C in the circuit (please show your work)?

1.2) (10 points) What is the tolerance of C (i.e. Cmin and Cmax) to meet 500Mbps data rate if all other circuit parameters are assumed to be ideal (no variation)? (Please show your work)

1.3) (10 points) If the tolerance of C is +/-10pF and assume all other parameters are ideal, what is the data rate limit (i.e. maximum data rate) for such I/O circuit (please show your work)? (slightly tricky!)

Problem 2 (VLSI buffer circuit design techniques) The delay versus fanout of a specific VLSI buffer circuit is shown in figure 2.1. Please answer the following questions.

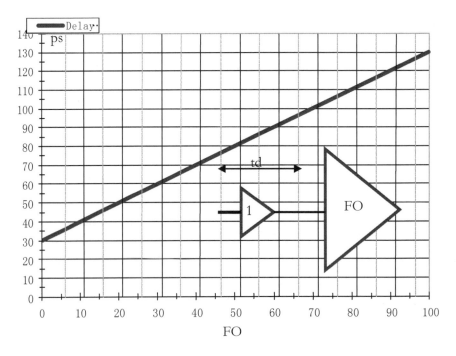

Fig.2.1 simulated delay versus FO curve

2.1) (10 points) Calculate parameter to and γ of such buffer circuit (Hin: to$(\gamma+FO)/(\gamma+1)$)

n	$(100)^{(1/n)}$
1	100.0
2	10.0
3	4.6
4	3.2
5	2.5
6	2.2
7	1.9
8	1.8
9	1.7
10	1.6

2.2) (7 points) For a capacitor load equals to FO = 100, how many buffer stages (include the first stage) should be used to achieve the minimal delay? (8 points) How small can this delay be?

Problem 3. (Quick questions). Please **circle** the best answer to each question (only **ONE** selection per question please!). **Don't spend too much time on an individual question**. It is OK to guess the answer if you are not very sure about the correct answer to that question.

Problem 3.1-3.6) For the T-line circuit show below, answer the following questions

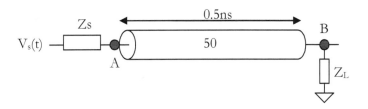

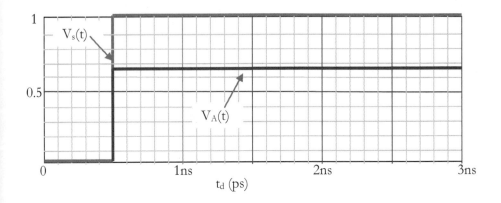

3.1) (3 points) What is likely the impedance Z_s based on above waveforms?

A) 25Ω   B) 50Ω   C) 75Ω   D) 100Ω

3.2) (3 points) What is likely the impedance Z_L based on above waveforms?

A) 25Ω   B) 50Ω   C) 75Ω   D) 100Ω

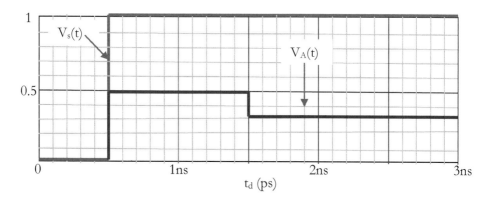

3.3) (3 points) What is likely the impedance Z_s based on above waveforms?

A) 25Ω   B) 50Ω   C) 75Ω   D) 100Ω

3.4) (3 points) What is likely the impedance Z_L based on above waveforms?

A) 25Ω   B) 50Ω   C) 75Ω   D) 100Ω

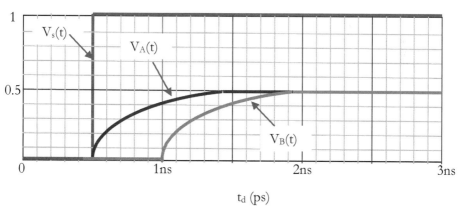

3.5) (3 points) Which of the following statements is likely true in the waveform above? (slightly tricky one!)

    A)   Zs consists of a capacitor in series with a resistor
    B)   Zl consists of a capacitor in series with a resistor
    C)   There is a capacitor between node A and ground
    D)   There is a capacitor between node B and ground

3.6) (3 points) What is likely this capacitor value based on above waveforms? (tricky one!)

A) 5pF   B) 10pF   C) 20pF   D) 30pF

3.7-3.8) A VLSI frequency divider circuit employing identical divide-by-2 circuit structures is shown below. Please answer the following questions:

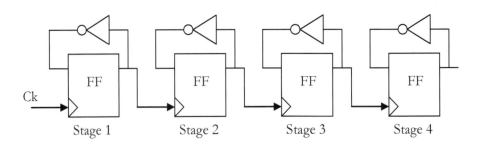

3.7) (3 points) Which stage is likely to fail first if there a setup time violation?

A) stage 1   B) stage 2
C) stage 3   D) stage 4   E) all stages are equally likely

3.8) Which stage is likely to fail first if there a hold time violation?

A) stage 1   B) stage 2
C) stage 3   D) stage 4   E) all stages are equally likely

3.9-3.10) A Bathtub curve of a high-speed I/O circuit is shown below. Please answer the following questions.

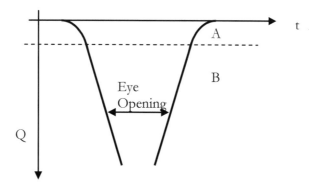

3.9) (3 points) Which of the following statements is most likely true?

A) Region A is dominated by DJ

B) Region A is dominated by RJ

C) Region A is dominated by AJ

D) Region A is dominated by CCJ

3.10) (3 points) Which of the following statements is most likely true?

A) Region B is dominated by DJ

B) Region B is dominated by RJ

C) Region B is dominated by AJ

D) Region B is dominated by CCJ

3.11 – 3.13) What are likely the best answers to the following questions?

3.11) (3 points) Power supply noise is a major contributor of _____.

A) RJ   B) ISI   C) Periodic Jitter (PJ)   D) CCJ

3.12) (3 points) Channel Bandwidth Limitation effect is a major contributor of
____.

A) RJ   B) ISI   C) Periodic Jitter (PJ)   D) CCJ

3.13) (3 points) Major source of RJ is ____.

A) Cross coupling noise effects   B) Channel reflection and ring effects
C) Channel bandwidth limitation effects   D) Thermal noise effects

3.14 –3.15) A VLSI D-FF is constructed from a basic VLSI D-FF as shown in
figure below. Please answer the following questions.

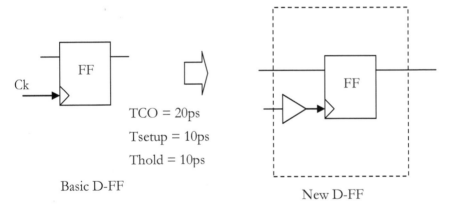

Ck

TCO = 20ps
Tsetup = 10ps
Thold = 10ps

Basic D-FF

New D-FF

3.14) (3 points) The setup time of the new D-FF is likely to be ____.

A)   0ps   B)   10ps
C)   20ps   D)   30ps

3.15) (3 points) The hold time of the new D-FF is likely to be ___.

A)   0ps   B) 10ps

C)   20ps   D) 30ps

Midterm Exam 2

EEE598 VLSI High-Speed I/O Circuits

Fall 2010

Arizona State University

Instructor: Dr. Hongjiang Song

Exam Time: 6:00pm – 7:15pm

Tuesday, Nov. 2, 2010

Print Name: _____

Signature: _____

ID#: _____

General instruction: This is a closed book/notes exam. However, you may bring a piece of paper (8 x11) with useful notes. You may also bring a calculator to exam.

Hint:

Good luck!

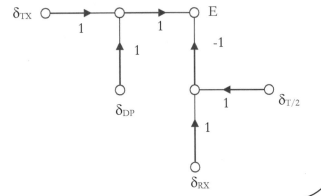

Problem 1 (Prototype I/O circuit model, I/O timing equation, VLSI PI circuit) A 5Gbps (i.e. T = 200ps) VLSI high-speed I/O circuit is shown in figure below. Such I/O circuit employs a <u>VLSI linear-based (i.e. $I_1+I_2+I_3+I_4 = IB$) PI circuit for optimal data sampling.</u> For simplicity, please ignore the setup and hold times of the Rx. We also ignore the parasitic capacitances effects of the Tx and Rx clock circuits. <u>Assume $t_{PI} = 100ps$ when $I_1=I_B$ and $I_2=I_3=I_4=0$.</u>

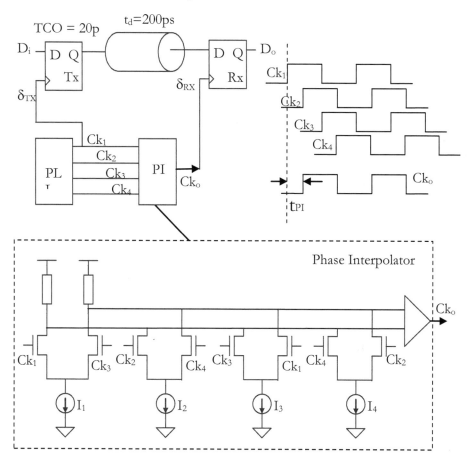

Fig.1.1 Conceptual VLSI PI based high-speed I/O circuit

1.1)   (10 points) For optimal circuit operation (i.e. maximum timing margin), what should be the value for t_{PI} in this high-speed I/O circuit (please show your work. You may use the back side of the cover page for more space)?

1.2)   (15 points) What should be the current bias values in terms of I_B ($I_1=?$, $I_2=?$, $I_3=?$, $I_4=?$) (Please show your work)

Problem 2. A VLSI programmable (i.e. multi-mode) frequency divider circuit with state output [Q1, Q2, Q3] is shown in figure below. The operation modes of the circuit can be set by control pin S.

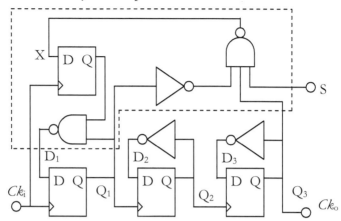

2.1) (15 points) What is the division ratio for <u>operation mode for S = 0</u>? (Please show your work by filling the necessary info in the given table)

CKi	Q1	Q2	Q3	QX	D1	D2	D3	X
1	0	0	0	1				
2								
3								
4								
5								
6								
7								
8								
9								
10								
11								
12								
13								
14								
15								
16								

2.2) (15 points) What is the division ratio for <u>operation mode for S = 1</u>? (Please show your work by filling the necessary info in the given table)

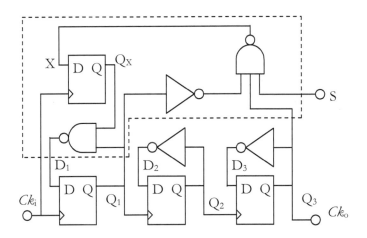

CKi	Q1	Q2	Q3	QX	D1	D2	D3	X
1	0	0	0	1				
2								
3								
4								
5								
6								
7								
8								
9								
10								
11								
12								
13								
14								
15								
16								

Problem 3. (Quick questions). Please **circle** the best answer to each question (only **ONE** selection per question please unless it is specified!). **Don't spend too much time on an individual question**. It is OK to guess the answer if you are not very sure about the correct answer to that question.

3.1 – 3.5) A VLSI PLL circuit is shown in figure below, where all delay elements are identical. Please answer the following questions. Assume the PLL is already locked-in and ignore the input capacitance of the frequency divider.

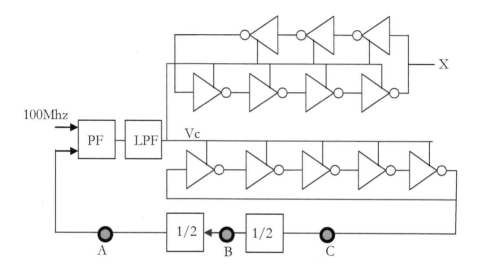

(3.1) (3 points) What is likely the clock frequency at <u>node A</u>?

a) 2Mhz   b) 100Mhz   c) 2.5Ghz   d) 5Ghz

(3.2) (3 points) What is likely the clock frequency at <u>node B</u>?

a) 2Mhz   b) 100Mhz   c) 2.5Ghz   d) 5Ghz

(3.3) (3 points) What is likely the clock frequency at <u>node C</u>?

a) 2Mhz   b) 100Mhz   c) 2.5Ghz   d) 5Ghz

(3.4) (3 points) What is likely the clock period at <u>node X</u>?

a) 100ps   b) 250ps   c) 280ps   d) 300ps

3.5) (3 points) *Signal Vc* is likely a

a) clock signal   b) slow varying signal

c) constant voltage signal   d) pulse signal

3.6 – 3.8) Two VLSI clock phase detector (CPD) circuits are shown in figure below, where the switches (SW1 and SW2) serve as "short" (i.e. R=0) when the control voltage is Vcc and as "open" (i.e. R = infinite) when the control voltage is Vss (= 0). Assume the R*C is much higher than the period of the input clocks. Please answer the following questions for the <u>steady state conditions</u> (i.e. t → infinite).

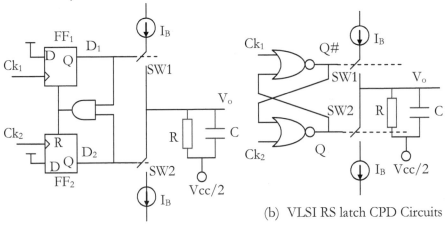

(a) VLSI PFD Circuits

(b) VLSI RS latch CPD Circuits

(3.6) (3 points) If Ck1 is a 200Mhz clock and Ck2 is a 100Mhz clock, what is likely the steady-state output voltage Vo for the VLSI PFD circuit?

a) 0   b) Vcc/2   c) Vcc/2+R*IB/2   d) Vcc

(3.7) (3 points) If Ck1 is a 200Mhz clock and Ck2 is a 100Mhz clock, what is likely the steady-state output voltage Vo for the VLSI RS latch CPD circuit?

a) 0   b) Vcc/2   c) Vcc/2+R*IB/2   d) Vcc

(3.8) (3 points) If both Ck1 and Ck2 are 100Mhz clocks and Ck1 leads Ck2 by T/2, what is likely the steady-state output voltage Vo for the VLSI PFD circuit?

a) 0   b) Vcc/2   c) Vcc/2+R*IB/2   d) Vcc

3.9 – 3.11) The SFG of a second order VLSI PLL circuit is shown in figure below. Please answer the following questions regarding to the noise transfer functions.

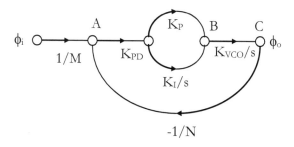

(3.9) (3 points) If noise is injected from point A to the PLL circuit, the noise transfer function will show a _____ frequency response.

a) ALL pass   b) Low-pass   c) High-Pass   d) Band-Pass

(3.10) (3 points) If noise is injected from <u>point B</u> to the PLL circuit, the noise transfer function will show a _____ frequency response.

a) ALL pass   b) Low-pass   c) High-Pass   d) Band-Pass

(3.11) (3 points) If noise is injected from <u>point C</u> to the PLL circuit, the noise transfer function will show a _____ frequency response.

a) ALL pass   b) Low-pass   c) High-Pass   d) Band-Pass

_____

(3.12) For the VLSI frequency divider circuit below, D-FF X is added for the purpose of

a) reducing the power dissipation   b) changing the frequency division ratio

c) reducing jitter of the circuit   d) improving the circuit operation speed

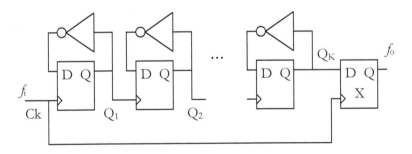

(3.13-3.15) A data recovery circuit that extract data signal phase to regenerate the Rx sampling clock (i.e. δ_{RX}) based on a digital control loop for optimal data sampling is shown in figure below. Please select <u>all applicable answers</u> (i.e. you may select multiple answers for the following questions) for the following questions based on the given VLSI phase detection (PD) implementations.

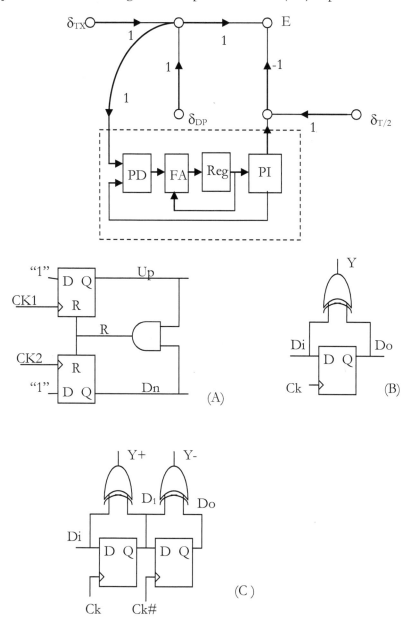

3.13) (3 points) What is (are) likely the problem(s) with the PD circuit <u>structure</u> <u>A</u>?

a) It is input clock duty-cycle dependent   b) It is input data pattern dependent

c) It cannot detect harmonics   d) Output of the PD is an analog signal type

3.14) (3 points) What is (are) likely the problem(s) with the PD circuit <u>structure</u> <u>B</u>?

a) It is input clock duty-cycle dependent   b) It is input data pattern dependent

c) It cannot detect harmonics   d) Output of the PD is an analog signal type

3.15) (3 points) What is (are) likely the problem(s) with the PD circuit <u>structure</u> <u>C</u>?

a) It is input clock duty-cycle dependent   b) It is input data pattern dependent

c) It cannot detect harmonics   d) Output of the PD is an analog signal type

Final Exam

EEE598 VLSI High-Speed I/O Circuits

Fall 2010
Arizona State University
Instructor: Dr. Hongjiang Song
Exam Time: 2:30pm – 4:20pm Thursday, Dec. 9, 2010

Print Name: _____

Signature: _____

ID#: _____

General instruction: This is a closed book/notes exam. However, you may bring three sheets of paper (8 x11) with useful notes. You may also bring a calculator to exam.

Good luck!

Problem1. A VLSI clock duty-cycle distortion (DCD) (e.g. non 50/50 duty-cycle) model is shown in figure below, where the DCD can be modeled using an added DC offset B on top of an ideal sinusoidal clock with respect to its desired common-mode value (Vcm = 0 in this model):

$$V(t) = B + A\sin(2\pi f_{ck}t)$$

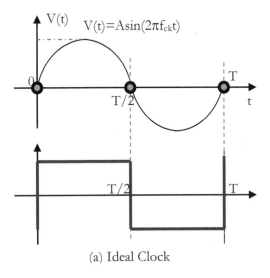

(a) Ideal Clock

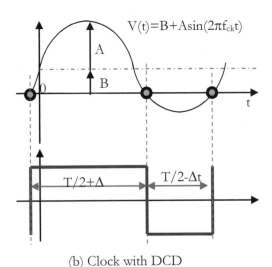

(b) Clock with DCD

1.1) (10 points) For small DCD (i.e. A>>B), find the expression of the DCD parameter $\underline{\Delta t}$ in terms of parameter A, B and clock period T (T =1/f_{ck}). (Hint $\sin(x) \sim x$ if $|x|<<1$)

1.2) (7 points) When such clock goes through a RC lowpass filter as show in figure 1.2, what is the gain of the DCD? (i.e. the ratio of Δt of the output clock versus input clock passing the lowpass filter)

$(\Delta t)_o/(\Delta t)_i = ?$

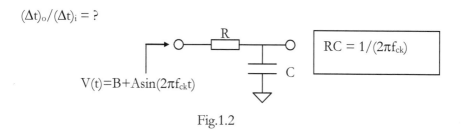

$$RC = 1/(2\pi f_{ck})$$

$V(t)=B+A\sin(2\pi f_{ck}t)$

Fig.1.2

1.3) (7 points) When such clock goes through a RC highpass filter as show in figure 1.3, what is the gain of the DCD? (i.e. the ratio of Δt of the output clock versus input clock passing the highpass filter).

$(\Delta t)_o/(\Delta t)_i = ?$

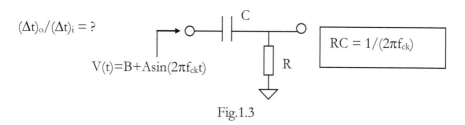

$$RC = 1/(2\pi f_{ck})$$

$V(t)=B+A\sin(2\pi f_{ck}t)$

Fig.1.3

Problem2. A first order s-domain SFG model of a 5Gps (i.e. T = 200ps) VLSI high-speed I/O data recovery circuit is shown in figure below. For simplicity we ignore the setup and hold time of the receiver and assume $\delta_{T/2} = \delta_{T/2}*$. Let $\omega_o/2\pi = 10$Mhz.

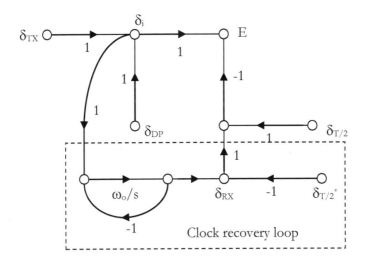

2.1) (10 points) Express δ_{Rx} and E as function of δ_i and (ω_o/s).

2.2) (5 points) For sinusoidal jitter at the Rx input (i.e. $\delta_i = (\delta_{tx} + \delta_{Dp}) = A\cos(\omega t)$) what is the peak value of E in terms of A at $f = \omega/2\pi = 10Mhz$?

2.3) (10 points) Find the jitter tolerance of this high-speed I/O receiver at 10Mhz jitter frequency. (Hint: The Rx jitter tolerance is define as the maximum allowed sinusoidal input jitter magnitude A (i.e. $\delta_i = (\delta_{tx} + \delta_{Dp}) = A\cos(\omega t)$) such that $E < E_{max}$ (i.e. no data transmission failure occur).)

Problem 3. (Quick questions) Please **circle** the best answer to each question (only **ONE** selection per question please!). **Don't spend too much time on an individual question**. It is OK to guess the answer.

3.1) (3 points) Three basic high-speed I/O circuit architectures are ___.

a) Common Clock I/O, Synchronous clock I/O, and embedded clock I/O.

b) Common Clock I/O, Synchronous clock I/O, and data recovery I/O.

c) Common Clock I/O, Source Synchronous I/O and embedded clock I/O.

d) Common Clock I/O, Synchronous clock I/O and Source Synchronous I/O.

3.2) (3 points) One major goal of forward clock I/O architecture is to minimize the phase tracking E in the I/O SFG model by ___.

a) minimizing the variation of δ_{Tx} term.

b) minimizing the variation of $\delta_{T/2}$ term.

c) minimizing the variation of TCO term.

d) minimizing the variation of t_D term.

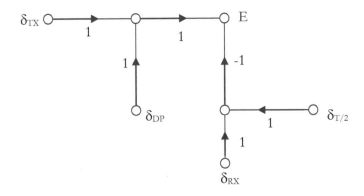

3.3) (3 points) A ½-frequency clock (i.e. 3Ghz clock for 6Gbps I/O) is commonly used in source synchronous I/O circuit for the purpose of____.

a)   Reducing the channel jitter amplification effects.

b)   Design simplicity.

c)   Low power

d)   Low cost

3.4) (3 points) A major advantage of the embedded I/O circuit versus the source synchronous I/O circuit is____.

a)   The reduced the channel jitter amplification effects.

b)   The design simplicity.

c)   The low power

d)   The low cost

3.5) (3 points) Channel coding techniques such as 8B/10B and 4B/5B coding are commonly used in VLSI high-speed I/O to ____.

a)   Improve the channel bandwidth

b)   Improve data transmission efficiency

c)   Reduce the power dissipation

d)   Ensure proper data recovery operation

3.6) (3 points) For the four types of filter-like circuit frequency responses shown in figure below, what is the correct order of circuits to have <u>increasing jitter amplification effects</u> (where fo is the clock frequency)

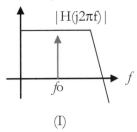

(I)

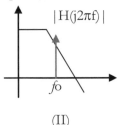

(II)

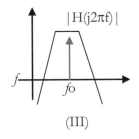

(III)

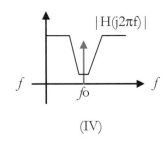

(IV)

a) (III)→(I)→(II)→(IV)

b) (III)→(IV)→(I)→(II)

c) (I)→(II)→(III)→(IV)

d) (IV)→(III)→(I)→(II)

3.7) (3 points) What is most likely the delay time of a 5-inch PCB trace?

a) 100ps
b) 1ns
c) 10ns
d) 100ns

3.8) (3 points) High-Speed I/O transmitter slew rate control circuit is used to ____.

a) Compensate for the channel bandwidth limiting effects
b) Reduce thermal noise effect
c) Minimize the reflection/ringing effect
d) Reduce the power dissipation

3.9) (3 points) High-Speed I/O transmitter slew rate control circuit is less effective at very high data rate because of____

    a)   Channel is too lossy to compensate

    b)   Crosstalk effect is too large to compensate

    c)   Channel bandwidth is too narrow to compensate

    d)   Signal band is too close to the resonant frequency of the channel to compensate

3.10) (3 points) What does <u>SIPO</u> in VLSI high-speed I/O circuit standard for?

    a)   Sign-In-Positive-Out

    b)   Single-In-Parallel-Out

    c)   Serial-In-Polarity-Out

    d)   Serial-In-Parallel-Out

3.11-3.14) The impulse response of a VLSI high-speed I/O channel is shown in figure below. Please select the best answer to the following questions

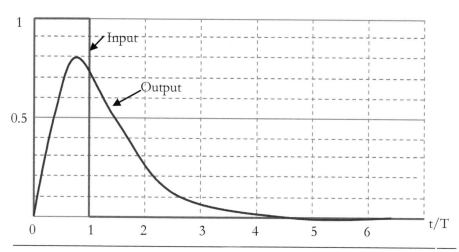

3.11) (3 points) Which of the following FIR filters is likely a good approximation of the <u>Channel</u>?

a)$H_{CH}(z) = 1 + 0.8Z^{-1} + 0.3Z^{-2}$

b) $H_{CH}(z) = 0.8 + 0.3Z^{-1} + 0.1Z^{-2}$

c)$H_{CH}(z) = 1 - 0.8Z^{-1} - 0.3Z^{-2}$

d) $H_{CH}(z) = 0.8 - 0.3Z^{-1} - 0.1Z^{-2}$

3.12) (3 points) Which of the following IIR filters is likely a good approximation of the <u>Equalizer</u> for this channel?

a)$H_{EQ}(z) = 1/(1 + 0.8Z^{-1} + 0.3Z^{-2})$

b) $H_{EQ}(z) = 1/(0.8 + 0.3Z^{-1} + 0.1Z^{-2})$

c)$H_{EQ}(z) = 1/(1 - 0.8Z^{-1} - 0.3Z^{-2})$

d) $H_{EQ}(z) = 1/(0.8 - 0.3Z^{-1} - 0.1Z^{-2})$

3.13) (3 points) Which of the following FIR filters is likely a good approximation of the <u>Equalizer</u> for this channel? (slightly trick!)

a)$H_{EQ}(z) = 1 + 0.8Z^{-1} + 0.3Z^{-2}$

b) $H_{EQ}(z) = 0.75 + 0.25Z^{-1}$

c)$H_{EQ}(z) = 1 - 0.8Z^{-1} - 0.3Z^{-2}$

d) $H_{EQ}(z) = 0.75 - 0.25Z^{-1}$

3.14) (3 points) What type of equalization circuit is commonly used in high-speed I/O Tx?

a)FIR   b) IIR   c) DFE   d) CTLE

3.15) (3 points) What does <u>DFE</u> in VLSI high-Speed I/O standard for?

    a) Digital feedback equalizer

    b) Deterministic feedback equalizer

    c) Differential feedback equalizer

    d) Decision feedback equalizer

3.16) For the clock waveform shown below, which of the statements best describes its signal power.

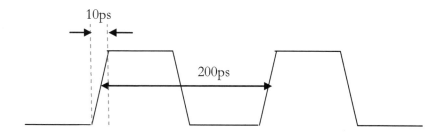

a) The signal power is approximately located within the frequency of 5Ghz;

b) The signal power is approximately located within the frequency of 2.5Ghz

c) The signal power is approximately located within the frequency of 50Ghz

d) The signal power is approximately located within the frequency of 100Ghz

3.17) An equivalent power delivery (PD) network of a VLSI circuit and its impedance profile are shown below, which of the expression is generally true?

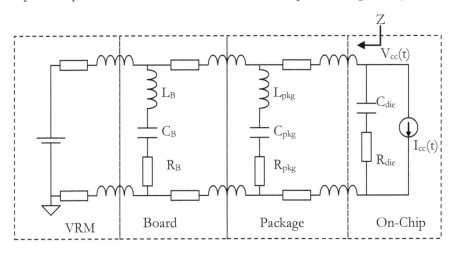

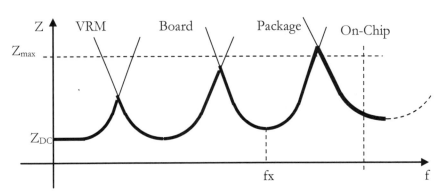

a) $fx \sim (L_{pkg}*C_{pkg})^{\wedge}(0.5)/(2\pi)$

b) $fx \sim (L_{pkg}*C_{pkg})^{\wedge}(-0.5)/(2\pi)$

c) $fx \sim (L_B*C_B)^{\wedge}(-0.5)/(2\pi)$

d) $fx \sim (L_{pkg}*C_{pkg}* L_B*C_B)^{\wedge}(-0.25)/(2\pi)$

Midterm Exam 1
EEE598: Serial Links
Fall 2012
Arizona State University
Instructor: Dr. Hongjiang Song
Exam Time: 6:00pm – 7:15pm Thursday, Sept. 27, 2012

Print Name: _____

Signature: _____

ID#: _____

General instruction: This is a closed book/notes exam. However, you may bring a sheet (8x11) of notes. You may also bring a calculator to exam.

Hint:

Good luck!

Problem 1 (VLSI VCDL) A VCDL circuit for DDR application is shown in below, where M0 and M1 are identical PMOS devices. All symmetrical load VCR are identical in this circuit with equivalent C and R (in term of 1/R versus the control voltage Vg) given in figure below. For simplicity you may assume ideal opamp and very fast CMOS technology with ignored parasitic capacitance (except the VCR).

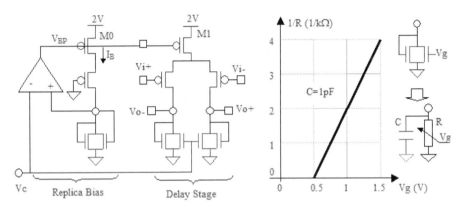

Fig.1 VLSI VCDL circuit

1.1) (6 points) Calculate the current bias I_B @ Vc = 1V.

1.2) (4 points) What is likely the threshold voltage V_T of the NMOS device in this process technology?

1.3) (10 points) If for a 533Mhz DDR2 input differential clock @ Vc =1V, find the V_{Max}, V_{Min} (i.e. low and High) values of the output clock signals.

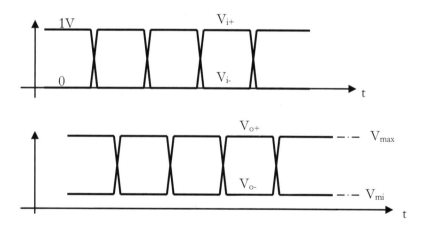

1.4) (10 points) Plot the delay time (in term of 1/td versus control voltage Vc) in the given graph for a Vc range of 0.5V < Vc <1.5V, and determine the delay time t_d @Vc = 1V.

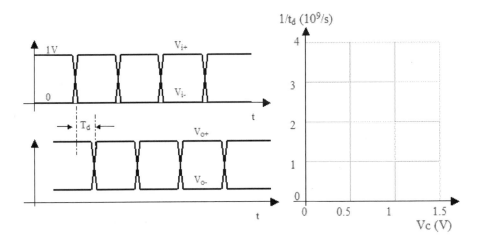

1.5) (4 points) Can this 533Mhz VCDL clock circuit to operate at V_C = 0.5V? (You only need to answer Yes? or No?)

Problem 2. (VLSI VCO circuits). A VCO for PCI-e application is implemented using 4 identical VCDL stages given in problem.1 with a shared replica bias shown in below.

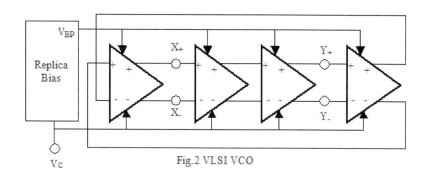

Fig.2 VLSI VCO

2.1) (15 points) For a given signal waveform at node X_+ below, plot the signal waveforms at nodes X_-, Y_+, and Y_- in the figure below.

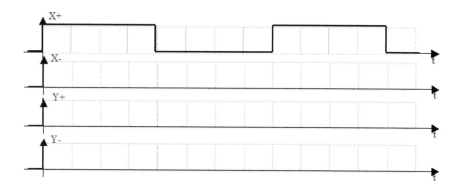

2.2) (10 points) Plot the frequency tuning curve (i.e. frequency versus Vc) for this VCO for 0.5V < Vc < 1.5V. Please label the scale of your frequency axis.

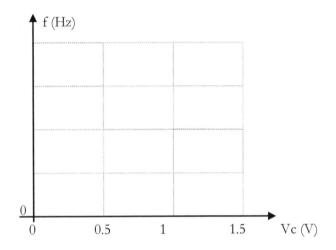

2.3) (5 points) What is the VCO gain K_{VCO} (where $KVCO = |df/d(Vc)|$) of this circuit?

Problem 3. (Common sense questions). Please **circle** the best answer (and only one!) to each question. **Don't spend too much time on an individual question.**

3.1-3.3) The timing SFG of a VLSI high-speed I/O is given below, where $\delta_{Tx} = 200$ps, $\delta_{DP} = 800$ps, $\delta_{T/2} = 100$ps. For simplicity, you may ignore the setup time and hold time (i.e. assuming 0) of the circuit.

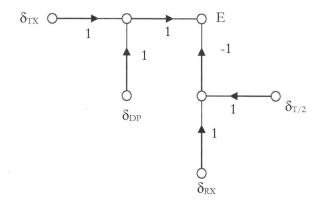

3.1) What is likely the targeted period of reference clock for this high-speed I/O?

a) 100ps   b) 200ps   c) 300ps   d) 400ps

3.2) What is likely the desired value for δ_{RX} for this VLSI high-speed I/O circuit?

a) 0ps   b) 50ps   c) 100ps   d) 200ps

3.3) What is the E_{max} for this I/O circuit?

a) 100ps   b) 200ps   c) 300ps   d) 400ps

3.4) Which of the following statements about the VLSI high-speed I/O clock eye opening in an eye diagram measurement is most likely true?

a) The eye usually increases with measurement time
b) The eye usually will be closed after a long measurement time
c) The eye usually is independent of the measurement time
d) The eye usually increases first, then it decreases

3.5) Which of the following statements about the I/O circuit jitter is most likely true?

a) DJ is usually unpredictable
b) DJ will always grow with measurement time
c) RMS value of RJ grows with measurement time
d) Peak vale of RJ grows with measurement time

3.6-3.8) The following are some basic VLSI interconnect design questions, where A is a uniformly distributed RC interconnect segment that has delay time (also known as flight time) td.

3.6) (3 points) What is likely the total delay time of the following interconnect structure?

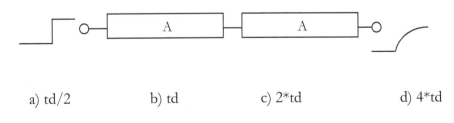

a) td/2   b) td   c) 2*td   d) 4*td

3.7) (3 points) What is likely the total delay time of the following interconnect structure?

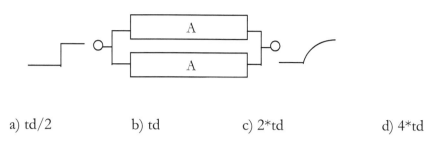

a) td/2   b) td   c) 2*td   d) 4*td

3.8) (3 points) If the optimal minimal delay repeater size (in term of gate capacitance) for interconnect A segments in serial is Cg, what is likely the optimal repeater size for two parallel interconnect paths of As as shown below?

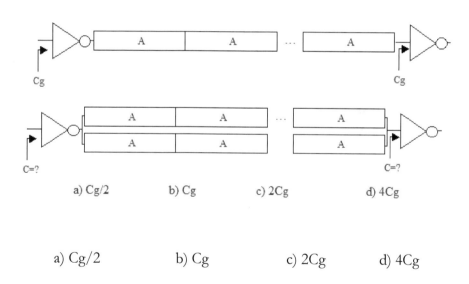

a) Cg/2   b) Cg   c) 2Cg   d) 4Cg

a) Cg/2   b) Cg   c) 2Cg   d) 4Cg

3.9-3.12) Assume a lossless T-line segment A has 50Ω T-lines impedance (Zo) and 1ns propagation delay (td) and assume there is no mute inductance coupling between the T-line segments when they are put together.

3.9) (3 points) What is likely the effective T-line impedance of the following T-line consisting of two A segments in series?

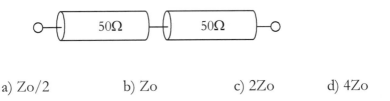

a) Zo/2   b) Zo   c) 2Zo   d) 4Zo

3.10) (3 points) What is likely the effective T-line impedance of the following T-line consisting of two A segments in parallel?

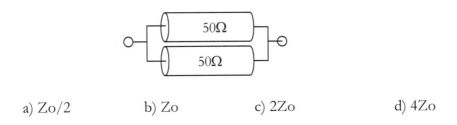

a) Zo/2   b) Zo   c) 2Zo   d) 4Zo

3.11) (3 points) What is likely the effective propagation delay time of the following T-line?

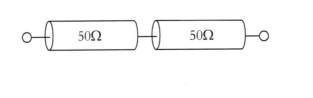

a) td/2   b) td   c) 2td   d) 4td

3.12) (3 points) What is likely the effective propagation delay time of the following T-line?

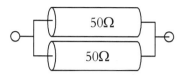

a) td/2   b) td   c) 2td   d) 4td

Midterm Exam 2

EEE598 VLSI High-Speed I/O Circuits

Fall 2012

Arizona State University

Instructor: Dr. Hongjiang Song

Exam Time: 6:00pm – 7:15pm Tuesday, Nov. 6, 2012

Print Name: _____

Signature: _____

ID#: _____

General instruction: This is a closed book/notes exam. However, you may bring a piece of paper (8 x11) with useful notes. You may also bring a calculator to exam.

Hint:

Good luck!

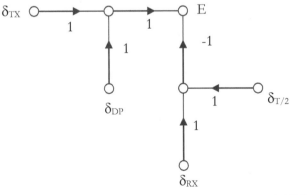

Problem 1 (VLSI PLL circuit analysis and design) A normalized (i.e. PLL circuit prototype using frequency scaling technique– you don't need to fully understand this technique for this problem) VLSI PLL circuit is shown in Figure below.

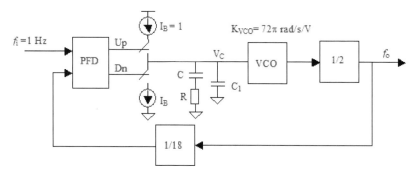

1.1)   (12 points) Determine parameter X, Y, Z, and K in the following SFG based on the circuit parameters (e.g. R, C, etc.) given above.

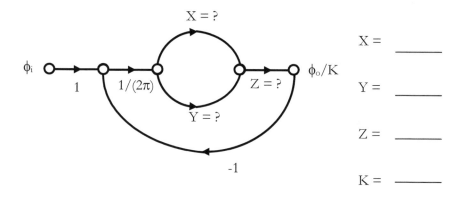

$X =$ _____

$Y =$ _____

$Z =$ _____

$K =$ _____

1.2)   (10 points) Derive the s-domain phase transfer function of the above PLL. (i.e. $\phi_o(s)/\phi_i(s) =?$) (please show your work)

1.3) (5 points) Determine the PLL output clock frequency f_o.

1.4) (15 points) Determine the circuit parameter C, C1, and R for the PLL for 1/10 Hz bandwidth and 0.7 Q-factor.

1.5) (5 points) Determine the divide-by-18 circuit for this PLL with output 50% duty-cycle based on existing logic gates and Flip-flops.

1.6)   (5 points) Determine the PFD circuit of the PLL based on existing logic gates and Flip-flops.

1.7)   (8 points) Determine the VLSI VCO circuit structure using MOS devices and ideal Op-amp that can be used for above PLL (please show your schematic).

1.8) (5 points) For the VCO with tuning curve given below and (0.4V – 1.0V) V_C operation range, calculate the Vc voltage at the PLL in the lock-in condition.

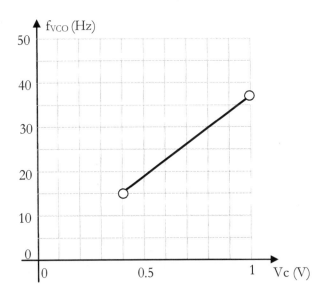

1.9) (5 points) If the two inputs of the PFD is swapped (from the original connection), what is likely the PLL output frequency? (slightly trickier)

Problem 2. (Quick questions). Please **circle** the best (and only one) answer to each question. **Don't spend too much time on an individual problem**.

2.1 – 2.3) For a VLSI synchronization circuit employing the basic bi-stable circuit structure shown below (where the PMOS and NMOS are balanced such that point C is in middle of point A and B). Assume the switch is closed (with low resistance R_{sw}) when CK is high and open (R = infinite) when CK is low. Let C be the total parasitic capacitance at node X and Y.

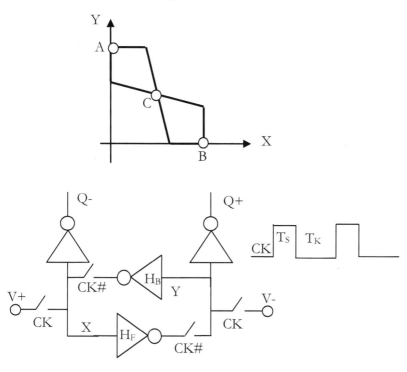

2.1) (3 points) For V+ >> V- in sampling phase (i.e. CK =”1”), what likely is the <u>final state</u> of the circuit for a long keep phase (i.e. CK = “0”)?

A)   Point A   B) Point B

B)   C) Point C   D)Somewhere between A and B

2.2) (3 Points) Which of the following statements about the minimum clock pulse Ts (i.e. time for CK ="1" phase) is likely true?

A)   Ts ~ $R_{sw}C$ B) Ts ~1/($R_{sw}C$)

C)   Ts independent of Rsw   D) Ts independent of C

2.3) (3 points) Which of the following statements about the minimum clock keep time TK (i.e. in CK ="0" phase) is likely true if V+>V- and $\Delta V = V_+ - V-$ is very small?

A) T_K ~ ΔV B) T_K ~ $1/\Delta V$

B)   C) T_K ~ $\log(1/\Delta V)$ D) TK independent of ΔV

2.4) (3 points) What does the PFD in PLL circuit stand for?

A)   phase forward detector   B) phase feedback detector

C)   phase frequency detector   D) phase feedback divider

2.5) (3 points) Which of the following statements is most likely true?

A) On-die termination circuit is used to control the slew rate of I/O circuit;

B) On-die termination circuit is used to improve the bandwidth of the I/O circuit;

C) On-die termination circuit is used to control the signal swing of the I/O circuit;

D) On-die termination circuit is used to minimize the jitter of the I/O circuit.

2.6) (3 points) Which of the following statements is most likely true?

A) A VLSI DLL circuit is commonly used to generate a high frequency clock from an external low frequency clock reference;

B) A VLSI PLL circuit has a Lowpass frequency response for power supply induce noise;

C) A PFD is usually used to eliminate harmonic (i.e. multi-cycle) lock-in of a VLSI DLL circuit;

D) A wide bandwidth is usually used in VLSI PLL to reject noise in the VCO circuit

2.7) (3 points) Which of the following statements about the PLL output jitter is most likely true?

A) PLL output jitter due to PFD noise usually increases as frequency increases;

B) PLL output jitter due to PFD noise usually first increases as frequency increases and then decrease with frequency;

C) PLL output jitter due to PFD noise usually decreases as frequency increases;

D) PLL output jitter due to PFD noise usually is independent of frequency

2.8) (3 points) Which of the following statements about the PLL output jitter is most likely true?

A) PLL output jitter due to LPF noise usually increases as frequency increases;

B) PLL output jitter due to LPF noise usually first increases and then decrease as frequency increases;

C) PLL output jitter due to LPF noise usually decreases as frequency increases;

D) PLL output jitter due to LPF noise usually is independent of frequency

2.9) (3 points) Which of the following statements about the PLL output jitter is most likely true?

A) PLL output jitter due to VCO noise usually increases as frequency increases;

B) PLL output jitter due to VCO noise usually first increases and then decrease as frequency increases;

C) PLL output jitter due to VCO noise usually decreases as frequency increases;

D) PLL output jitter due to VCO noise usually is independent of frequency;

2.10) (3 points) VLSI CPD is not suitable for detecting the data phase of high-speed I/O due to the facts that

A) It is too fast;

B) It is too slow:

C) It cannot detect harmonics in data signal;

D) It cannot handle random data pattern.

Final Exam

EEE598 Serial Links
Fall 2012
Arizona State University
Instructor: Dr. Hongjiang Song
Exam Time: 4:50pm – 6:40pm Thursday, Dec. 13, 2012

Print Name: _____

Signature: _____

ID#: _____

General instruction: This is a closed book/notes exam. However, you may bring three sheets of paper (8x11) with useful notes. You may also use calculator in the exam.

Good luck!

Problem1. A simplified VLSI high-speed I/O circuit including the TX, the channel and the RX is shown in figure below. (1) Assume the output impedances of the buffers are zero and ignore the delay of the buffers and the FFs, (2) Assume the circuits are powered by +/-0.5V supplies, and (3) assume the channel is an ideal transmission line of 50Ω impedance and 2ns delay.

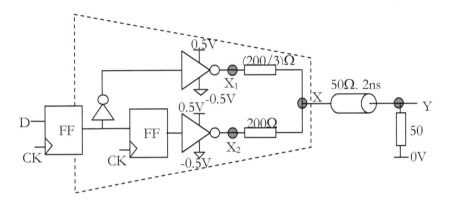

1.1)   (20 points) Draw waveforms at node X1, X2, X and Y for the given D and CK below:

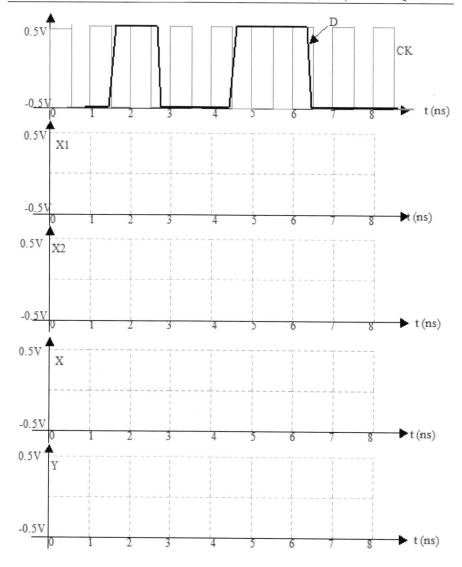

1.2) (5 points) Find the effective output impedance of the TX.

1.3) (10 points) Derive the Z-domain transfer function of the Tx and determine the De-emphasis level (dB) of the equalizer in the TX:

$$H_{Tx}(z) = X(z)/(D(z) = ?$$

De-emphasis (dB) = $20*LOG(|H_{tx}|_{f=fck/2}/|H_{Tx}|_{f=0})$ (i.e. ratio of Tx gain at $f_{ck}/2$ to DC gain) =?

Problem 2. (Quick questions) Please **circle** the best answer (only one) to each question.

2.1-2.4) For the I/O SFG model shown below, assume the Tx TCO is 10ps, the Rx sampler setup and hold times are both 5ps. The channel delay time across 1 million parts has a Gaussian distribution with $(td)_{min}$ = 1.8ns and $(td)_{max}$ = 2.2ns. We can also assume $\delta_{T/2} = \delta_{T/2}*$.

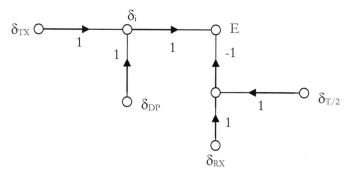

2.1) (3 points) What is likely the E_{max} for this link at 5Gps data rate?

a)   95ps   b) 200ps   c) 1.8ns   d) 2.2ns

2.2) (3 points) Assume a data recovery circuit as shown in below is used to achieve the 5Gbps data rate. What is likely the input phase error transfer function for this I/O circuit?

b) $E/\delta i = (s-\omega_o)/(s+\omega_o)$ b) $E/\delta i = s/(s+\omega_o)$

c) $E/\delta i = \omega_o/(s+\omega_o)$ d) $E/\delta i = s^2/(s+\omega_o)$

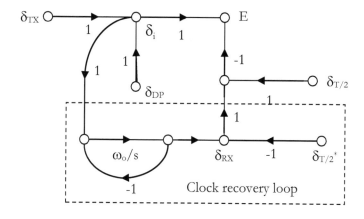

2.3) (3 points) For above embedded clock I/O circuit, what is likely the clock recovery transfer function ($\delta_{Rx}/\delta i$ @ $\delta_{T/2}^* =0$)?

a) $\delta_{Rx}/\delta i = s/(s+\omega_o)$

b) $\delta_{Rx}/\delta i = \omega_o/(s+\omega_o)$

c) $\delta_{Rx}/\delta i = (s-\omega_o)/(s+\omega_o)$

d) $\delta_{Rx}/\delta i = s^2/(s+\omega_o)$

2.4) (3 points) If $\delta_{TX} =0ps$, what is most likely the targeted Rx clock phase for designing this high-speed I/O at nominal condition? (assuming 5Gbps data rate)

a) 1.8ns   b) 110ps   c) 100ps   d) 10ps

2.5) (3 points) Which of the following VLSI circuit types can provide jitter attenuation function?

a) A lumped passive RC circuit

b) A typical VLSI channel

c) A distributed RC circuit (VLSI interconnect)

d) A RLC circuit

2.6) (3 points) When a VLSI high frequency clock is buffered using multiple stages of VLSI CMOS inverters, the clock duty-cycle error usually

a) Gets smaller from stage to stage along the clock distribution circuit.

b) Stays the same.

c) Becomes worse and worse.

d) Becomes smaller in the first few stages and then become worse.

2.7) (3 points) For a 10 inches length PCB trace, what is likely the minimum rise/fall time of the data signal that can be sent without requiring termination?

a) 10ps

b) 100ps

c) 1ns

d) 10ns

2.8) (3 points) Which of the following statements about the slew rate control in high-speed I/O is most likely true?

a) It is equivalent to increase the bandwidth of the I/O channel

b) It is equivalent to reduce the Q-factor of the I/O channel

c) It is equivalent to apply band-limiting operation to the signal

d) It is equivalent to minimize the xtalk of the I/O circuits.

2.9) (3 points) Which of the following statement about the termination of high-speed I/O is most likely true?

a) It is equivalent to increase the bandwidth of the I/O channel

b) It is equivalent to reduce the Q-factor of the I/O channel

c) It is equivalent to apply band-limiting operation to the signal

d) It is equivalent to minimize the xtalk of the I/O circuits.

2.10) Which of the following statement about the equalization of high-speed I/O is most likely true?

a) It is equivalent to increase the bandwidth of the I/O channel

b) It is equivalent to reduce the Q-factor of the I/O channel

c) It is equivalent to apply band-limiting operation to the signal

d) It is equivalent to minimize the xtalk of the I/O circuits.

2.11-2.15) When a VLSI PCB based I/O channel is measured in lab, the following waveform shown below is observed at node B.

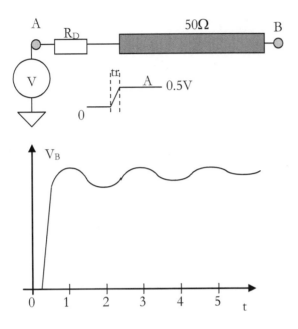

2.11) (3 points) Which of the following statements regarding to the driver impedance Rd is most likely true?

a) Rd = 50 Ω

b) Rd = 0 Ω

c) Rd >50Ω

d) 0<Rd<50Ω

2.12) (3 points) What is likely the length of this PCB trace?

a) 1 inch

b) 5 inch

c) 10 inch

d) 20 inch

2.13) (3 points) What is the most likely the minimum rise time (tr) at point A such that ringing at point B can be avoided?

a) 100ps

b) 1ns

c) 10ns

d) 100ns

2.14) (3 points) For 550ps rise time, what is likely the PCB trace requirement to avoid the ringing?

a) L< 0.5 inch   b) L < 5 inches   c) L < 10 inches   d) L < 20 inches

2.15) (3 points) Which the following power spectrum plots of a VLSI clock is most likely related to the effect of the DCD?

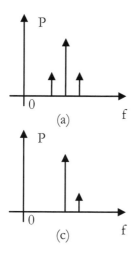

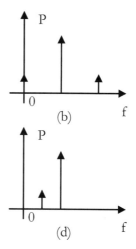

2.16) (3 points) Which of the following circuit blocks are usually **not** part of VLSI high-speed I/O **Tx** circuit?

a) Equalizer

b) SIPO

c) Termination

d) PISO

2.17) (3 points) Which of the following circuit blocks are usually **not** part of VLSI high-speed I/O **Rx** circuit?

a) Equalizer

b) SIPO

c) Termination

d) PISO

2.18) (3points) Which of the following circuit blocks are usually **not** part of VLSI high-speed I/O **DRC** circuit?

a) CPD

b) PI

c) VCO

d) VCDL

2.19) (3 points) One of the purposes for using 8B/10B encoding in the high-speed I/Os (such as PCI-E and SATA) is ____.

a) To reduce the power dissipation

b) To increase the channel bandwidth

c) To ensure the DRC loop operation

d) To reduce the xtalk effects

2.20 –2.22) Shown in figure is a typical impedance profile of a VLSI chip that shows three impedance peaks.

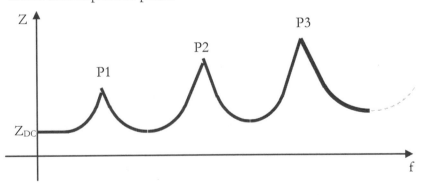

2.20) (3 points) Which of the following PD parameter will be impacted by changing the <u>on-chip decoupling capacitor?</u>

    a) Z_{DC}

    a) P1 and P2

    b) P2 and P3

    c) P3

2.21) (3 points) Which of the following PD parameter will be impacted by changing the <u>package decoupling capacitor?</u>

    a) Z_{DC}

    b) P1 and P2

    c) P2 and P3

    d) P3

2.22) (2 points) Which of the following PD parameter will be impacted by changing the <u>on-board decoupling capacitor?</u>

    a) Z_{DC}

    b) P1 and P2

    c) P2 and P3

    d) P3

VLSI CIRCUIT DESIGN SERIES

VLSI Analog Signal Processing Circuits – Algothrim, Architecture, Modeling, and Circuit Implementation. by Hongjiang Song, ISBN #978-1-4363-7740-9. (2009).

This is the textbook for the **VLSI Switched-Capacitor Filter and Analog Signal Processing Circuit Design** class (EEE598) the author offered at Arizona State Uinversity covering VLSI passive, active-RC, MOS-C, Gm-C, CTI, SC, SI analog filters and signal processing circuit techniques.

VLSI High-Speed I/O Circuits – Theoritical Basis, Architecture, Modeling, and Circuit Implementation. by Hongjiang Song, ISBN #978-1-4415-5987-6. (2010).

This is the textbook for the **VLSI high-speed I/O circuits** class (EEE598) the author offered in Arizona State University covering the analysis, modeling, and implementation of VLSI high-speed I/O circuits, such as timing models, jitter analysis, transmitter, receiver, equalizer, phase-locked loop (PLL), and data recovery circuit designs.

The Arts of VLSI Circuit Design – Symmetry Approaches Toward Zero PVT Sensitivities. by Hongjiang Song, ISBN #978-1-4568-7468-7. (2011).

This is the textbook for the **Structural VLSI Analog Circuit Design** class (EEE598) the author offered in Arizona State Uinversity covering various the state-of-the-arts symmetry based low PVT sensitivity circuit design techniques for basic VLSI circuit elements, circuit blocks and systems.

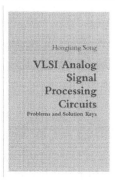

VLSI Analog Signal Processing Circuits – Problems and Solution Keys. by Hongjiang Song, ISBN #978-1-304-74949-9. (2013).

This book includes exam sheets and solution keys for the **Switched-Capacitor and Analog Signal Processing Circuit Design** class (EEE598) offered in Arizona State University in the past ten years covering VLSI passive, active-RC, MOS-C, Gm-C, CTI, SC and SI circuit techniques.

The Arts of VLSI Opamp Circuit Design by Hongjiang Song, ISBN #978-1-312-05130-0. (2014).

This book provides an introduction to structural VLSI opamp circuit design techniques and practices employing symmetry principles.

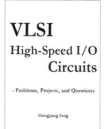

VLSI High-Speed I/O Circuits – Problems, Projects and Questions. by Hongjiang Song. (2014). ISBN # 978-1-312-05875-0

This book includes a collection of commonly seen class homework problems, design projects, and interview questions for VLSI high-speed I/O circuit design, modeling and implementations.

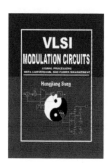

VLSI Modulation Circuits – Signal Processing, Data Conversion, and Power Managements. by Hongjiang Song. (to be published).

This book is based on the class notes the author developed for the School of Engineering at Arizona State University. The materials cover various design techniques of VLSI modulation circuits with applications in signal processing, data conversion, and power management.